Pixologic ZBrush 2018

for Novices

(Learn By Doing)

CADSoft Technologies

Pixologic ZBrush 2018 for Novices

CADSoft Technologies

ISBN 978-1-64057-077-1

NOTICE TO THE READER

Publisher does not warrant or guarantee any of the products described in the text or perform any independent analysis in connection with any of the product information contained in the text. Publisher does not assume, and expressly disclaims,any obligation to obtain and include information other than that provided to it by the manufacturer.

The reader is expressly warned to consider and adopt all safety precautions that might be indicated by the activities herein and to avoid all potential hazards. By following the instructions contained herein, the reader willingly assumes all risks in connection with such instructions.

The Publisher makes no representation or warranties of any kind, including but not limited to, the warranties of fitness for particular purpose or merchantability, nor are any such representations implied with respect to the material set forth herein, and the publisher takes no responsibility with respect to such material. The publisher shall not be liable for any special, consequential, or exemplary damages resulting, in whole or part, from the reader's use of, or reliance upon, this material.

DEDICATION

To students, who are dedicated to learning new technologies

THANKS

To employees of CADSoft Technologies
for their cooperation

Online Training Program Offered by CADSoft Technologies

CADSoft Technologies provides effective and affordable virtual online training on various software packages including Computer Aided Design, Manufacturing, and Engineering (CAD/CAM/CAE), computer programming languages, animation, architecture, and GIS. The training is delivered 'live' via Internet at any time, any place, and at any pace to individuals as well as the students of colleges, universities, and CAD/CAM/CAE training centers. The main features of this program are:

Training for Students and Companies in a Classroom Setting

A team of highly experienced instructors and qualified engineers at CADSoft Technologies conduct the classes in classroom environment at its centers. This team has authored several textbooks that are rated "one of the best" in their categories and are used in various colleges, universities, and training centers in North America, Europe, and in other parts of the world.

CADSoft Technologies with its cost effective and time saving initiative strives to deliver online training in the comfort of your home or work place, thereby relieving you from the hassles of traveling to training centers. CADSoft Technologies strives to be the best training institute in USA. CADSoft Technologies has developed career oriented programs by following state-of-the-art teaching and learning methodologies.

Training for Individuals

CADSoft Technologies with its cost effective and time saving initiative strives to deliver the training in the comfort of your home or work place, thereby relieving you from the hassles of traveling to training centers.

Training Offered on Software Packages

CADSoft provides basic and advanced training on the following software packages:

***CAD/CAM/CAE**: CATIA, Pro/ENGINEER Wildfire, Creo Parametric, Creo Direct, SOLIDWORKS, Autodesk Inventor, Solid Edge, NX, AutoCAD, AutoCAD LT, AutoCAD Plant 3D, Customizing AutoCAD, EdgeCAM, and ANSYS*

***Architecture and GIS**: Autodesk Revit (Architecture, Structure, MEP), AutoCAD Civil 3D, AutoCAD Map 3D, Navisworks, Oracle Primavera, and Bentley STAAD Pro*

***Animation and Styling**: Autodesk 3ds Max, Autodesk Maya, Blender, Pixologic ZBrush, and MAXON CINEMA 4D*

***Computer Programming**: C++, VB.NET, Oracle, AJAX, and Java*

***Personality Development**: Personality Development and Engineering Ethics/Soft Skills Course*

Table of Contents

Preface

Pixologic ZBrush 2018

Pixologic ZBrush 2018 is a powerful modeling and sculpting software developed by Pixologic Inc. This software is used for developing highly detailed characters for movies, games, and digital design projects.

Pixologic ZBrush 2018 for Novices textbook covers all features of ZBrush 2018 in a simple, lucid, and comprehensive manner. It gives in-depth details of the concepts and explains the usage and functions of ZBrush such as DynaMesh, NanoMesh, ZRemesher, ZModeler, NanoMesh, and KeyShot renderer. In this edition, new features such as Gizmo 3D and the Live Boolean mode, which is used to generate boolean results, have been explained. This textbook will unleash your creativity and transform your imagination into reality, thus helping you create realistic 3D models. This textbook caters to the needs of both the novice and advanced users of ZBrush 2018 and is ideally suited for learning at your convenience and at your pace.

The salient features of this textbook are as follows:

- **Tutorial Approach**
  The author has adopted the tutorial point-of-view and the learn-by-doing approach throughout the textbook. This approach will guide the users through the process of creating the models, adding textures to them, and animating them in the tutorials.

- **Real-World Models as Projects**
  The author has used about 24 real-world sculpting and modeling projects as tutorials in this textbook. This will enable the readers to relate the tutorials to the real-world models in the animation industry. In addition, there are about 20 exercises that are also based on the real-world modeling projects.

- **Tips and Notes**
  Additional information related to various topics is provided to the users in the form of tips and notes.

- **Learning Objectives**
  The first page of every chapter summarizes the topics that are covered in the chapter. This will help the users to easily refer to a topic.

- **Exercises**
  Exercises are given at the end of the chapter and they can be used by the instructors to assess the knowledge of the student. You need to perform all the exercises as output of some of these exercises may have been used as a reference later in this book.

Symbols Used in the Textbook

Note
The author has provided additional information to the users about the topic being discussed in the form of notes.

Tip
Special information and techniques are provided in the form of tips that helps in increasing the efficiency of the users.

Formatting Conventions Used in the Textbook

Please refer to the following list for the formatting conventions used in this textbook.

- Names of palettes, subpalettes, buttons, sliders, and areas are written in boldface. Example: The **Tool** palette, the **Geometry** subpalette, the **Load Tool** button, the **SDiv** slider, and the **DynaMesh** area.
- Names of dialog boxes and the check boxes are written in boldface. Example: The **Projection Master** dialog box, the **Fade** check box.
- Names of the files are italicized. Example: *c08tut3start.ZTL*

Naming Conventions Used in the Textbook

Flyout

A flyout is a menu that contains the tools having similar type of functions. Figure 1 shows the flyout displayed on choosing the Current Brush button.

Palettes

The palettes consist of tools and settings that are used to perform different tasks in ZBrush. Each palette is dedicated to a single set of related features. For instance, the **Color** palette consists of different buttons and sliders that are used to select solid and gradient colors for filling the canvas or an object. Figures 2 and 3 show the **Color** and **File** palettes, respectively.

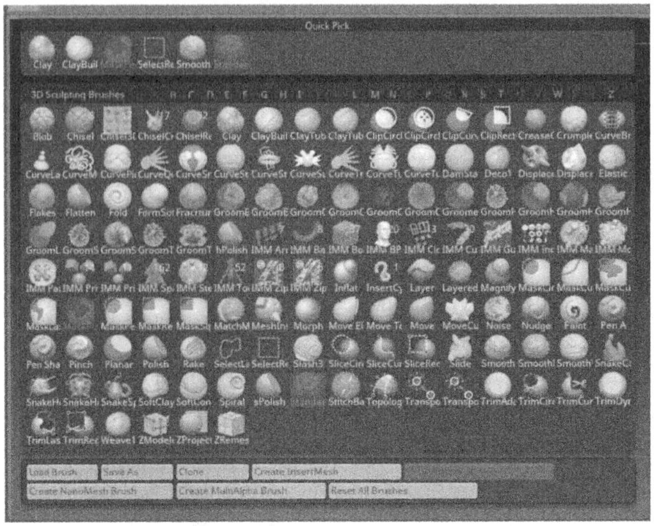

Figure 1 The flyout displayed on choosing the Current Brush button

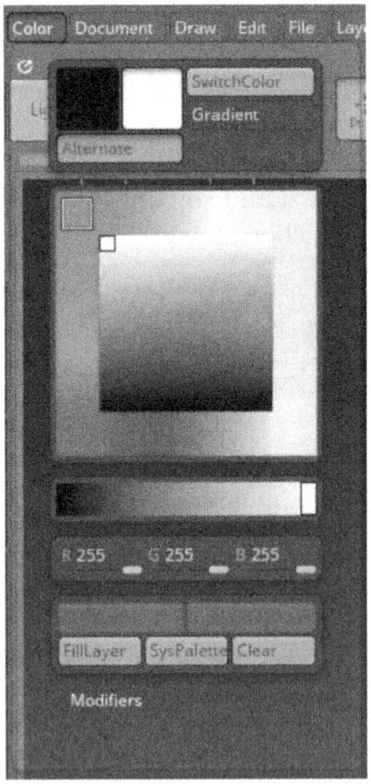

*Figure 2 The **Color** palette*

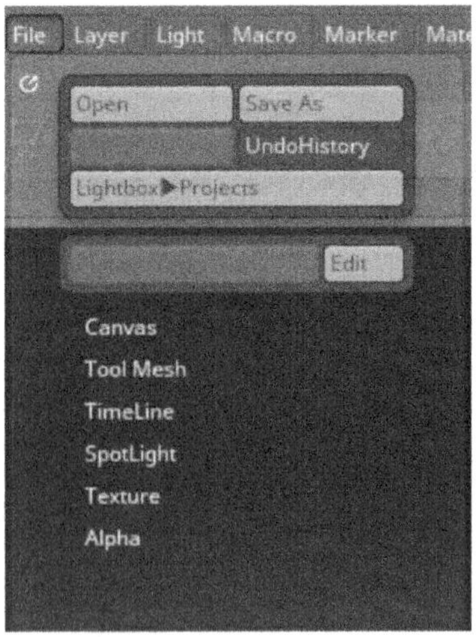

*Figure 3 The **File** palette*

Subpalettes

The subpalettes are the menus inside each palette containing different settings which are used for sculpting and modeling. Figure 4 shows the **Geometry** subpalette located in the **Tool** palette and Figure 5 shows the **UV Master** subpalette located in the **ZPlugin** palette.

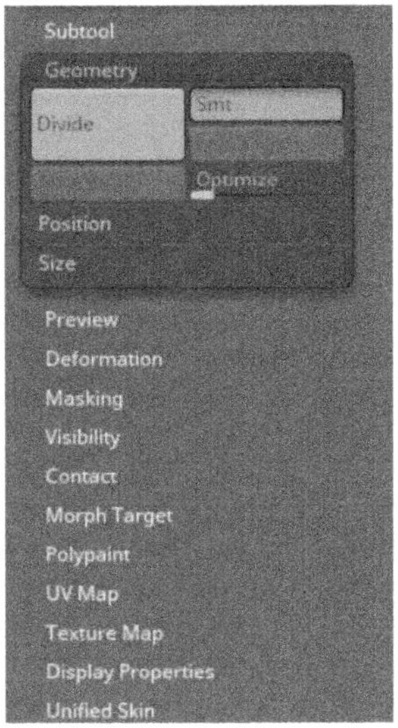

Figure 4 *The* ***Geometry*** *subpalette*

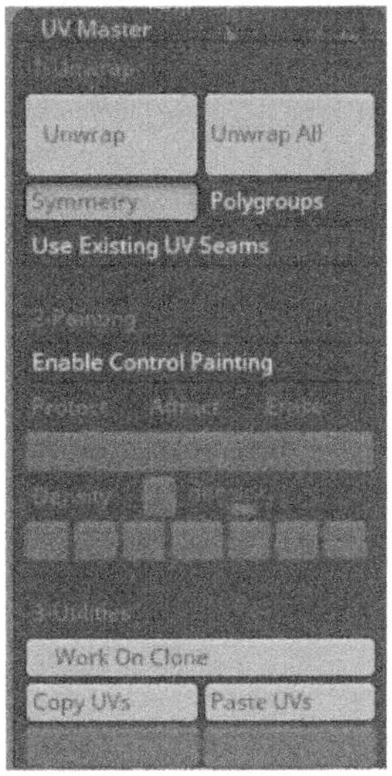

Figure 5 *The* ***UV Master*** *subpalette*

Page intentionally left blank

Chapter 1

Exploring ZBrush Interface

Learning Objectives

After completing this chapter, you will be able to:

- *Work with various tools available in the ZBrush interface*
- *Open and save a ZTool and ZDocument*

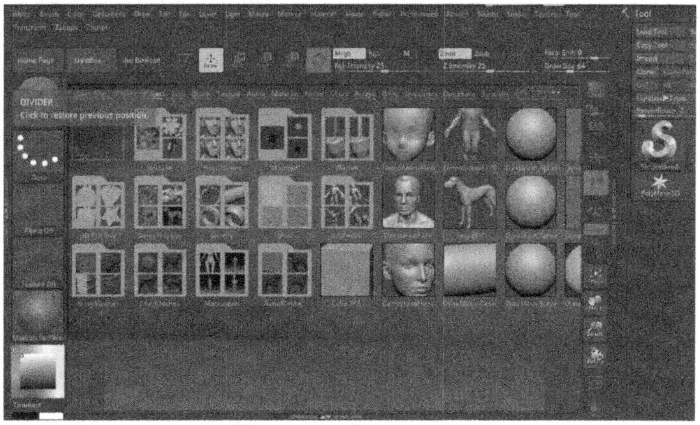

INTRODUCTION TO PIXOLOGIC ZBrush 2018

ZBrush is a modeling and sculpting software used to create realistic 3D models. It was developed by Pixologic Inc. in the year 1999. It has a unique blend of 2D, 2.5D, and 3D features. ZBrush consists of different tools that enable you to create or import 3D models, and then add high level details to them. This software enables you to sculpt on high resolution models that consist of millions of polygons. ZBrush is used by all the major animation studios worldwide. It has been used in the movies such as 300, Pirates of the Caribbean, Avatar, Hulk, and so on. ZBrush 2018 is compatible with different 3D software applications such as Maya, 3ds Max, CINEMA 4D, and Modo. Although, ZBrush is a vast software to deal with, yet all the major tools available in ZBrush 2018 are discussed in this textbook. In this chapter, you will learn about the ZBrush interface.

STARTING ZBrush 2018

To start ZBrush 2018, choose the **Start** button on the taskbar; the **Start** menu will be displayed. Next, choose **All Programs > Pixologic > ZBrush 2018 > ZBrush 2018** from the **Start** menu, as shown in Figure 1-1; the default ZBrush interface will be displayed with its various components along with the **ZBrush:Home Page** window, as shown in Figure 1-2 and Figure 1-3.

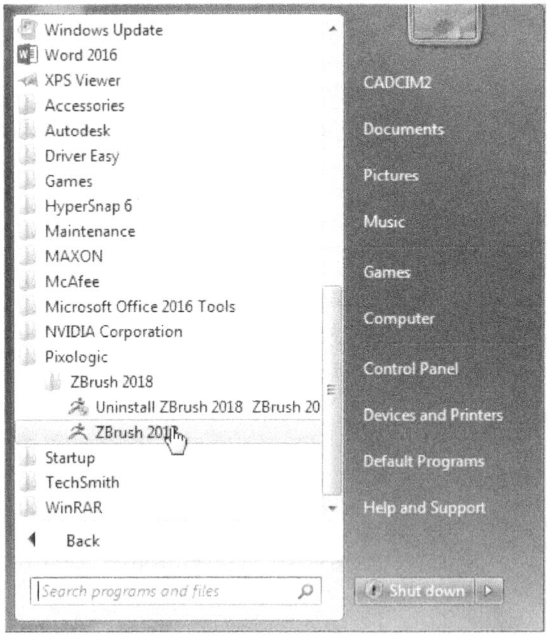

Figure 1-1 *Starting Pixologic ZBrush 2018 from the* ***Start*** *menu*

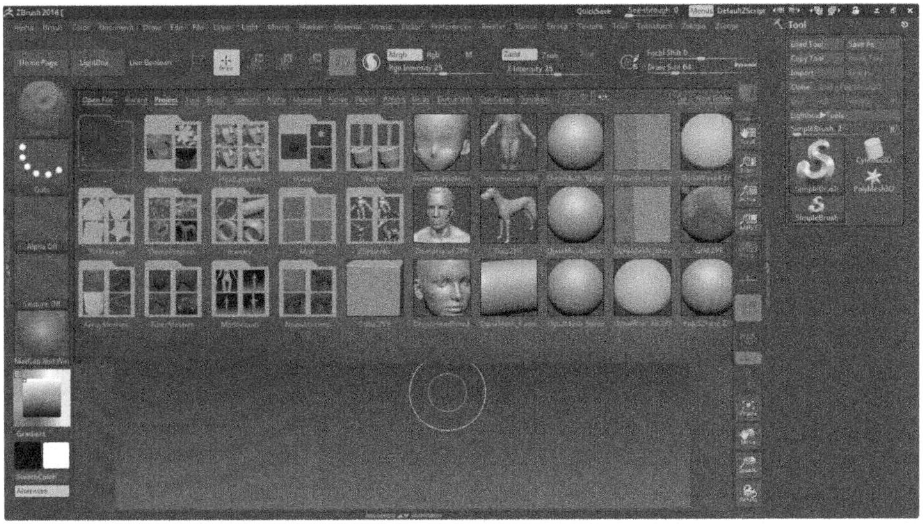

Figure 1-2 The default ZBrush 2018 interface

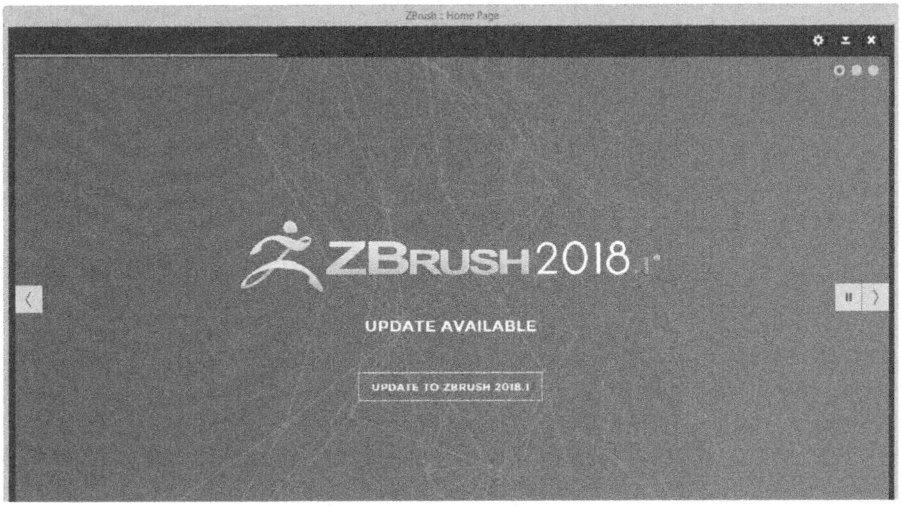

Figure 1-3 *The* ***Zbrush Home Page*** *window*

Alternatively, you can start ZBrush 2018 by double-clicking on its shortcut icon displayed on the desktop. This icon is automatically created on installing ZBrush 2018 on your computer.

EXPLORING THE ZBrush 2018 INTERFACE

ZBrush interface consists of components such as title bar, palettes, shelves, trays, and canvas. When you start ZBrush 2018 for the first time, the default interface will display various components of ZBrush 2018, refer to Figure 1-4. These components are discussed next.

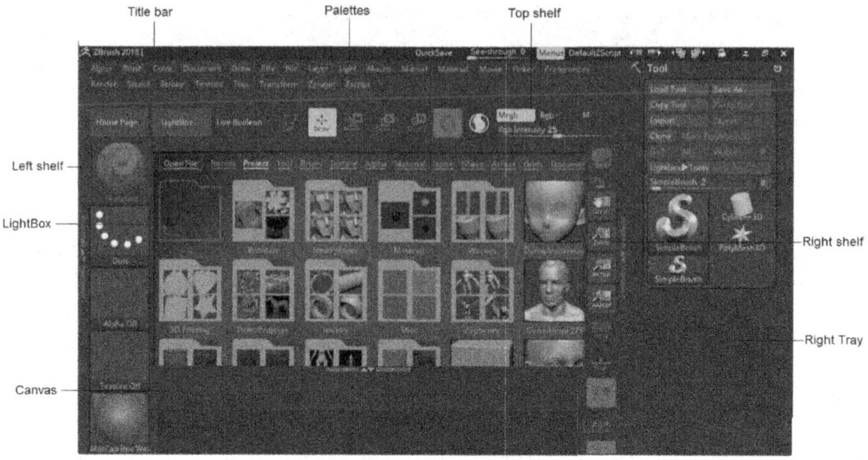

***Figure 1-4** Various components of the ZBrush 2018 interface displayed*

Title Bar

The title bar is located on the top of the ZBrush screen. In the title bar, various information such as version number, hardware ID, and memory usage is displayed on the left, refer to Figure 1-5. Besides this, the **QuickSave** button and the **See-through** slider are also present in the title bar. The **QuickSave** button is used to save the different stages of a file sequentially. These files can be accessed from the **QuickSave** tab of the **LightBox** browser. The **See-through** slider is used to increase or decrease the transparency in the ZBrush interface so that you can view the background screen. This slider makes it convenient for you to view the reference images in the background without importing a reference image or toggling back and forth between the image and interface. For example, if you are following the steps of a tutorial of ZBrush then you can view these steps in the background and follow the instructions with ease.

On the right side of the title bar, various buttons are displayed, as shown in Figure 1-6. The **Menus** button is used to toggle the display of the palettes. The **DefaultZScript** button is used to run the default scripting language present in ZBrush if it has been edited. The **Load Previous User Interface Colors**, **Load Next User Interface Colors** and **Load Previous User Interface Layout**, and **Load Next User Interface Layout** buttons are used to change the color and the layout of the screen, respectively. These presets enable you to customize the ZBrush screen.

***Figure 1-5** Partial view of the Title bar*

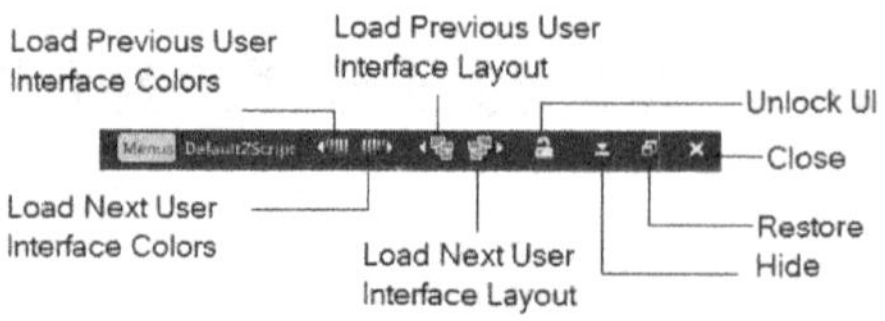

***Figure 1-6** Partial view of the title bar (right side)*

Canvas

The canvas is the drawing area where you can sculpt or create 3D models. It covers most of the area of the ZBrush interface, refer to Figure 1-4. The 3D models created can be saved as 2D illustrations. A 2D illustration is called ZBrush document and is saved in the ZBR format. The 3D models are called ZBrush tools and are saved in the ZTL format.

Palettes

Palettes are located below the Title bar. They are organized alphabetically, starting from the **Alpha** palette and ending at the **Zscript** palette, refer to Figure 1-7. These palettes are discussed next.

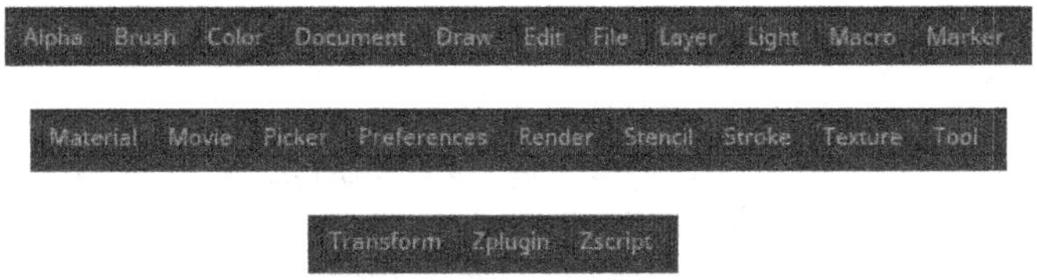

***Figure 1-7** The palettes*

Alpha

The **Alpha** palette consists of various gray scale images known as alphas. Alphas are available in different patterns. They determine the shape of the brush used in painting and sculpting. This palette also contains different subpalettes that are used to modify the alpha images as required.

Brush

The **Brush** palette consists of different types of 3D sculpting brushes. Besides this, it has different subpalettes that are used to modify the brush settings. The **Standard** brush is the default sculpting brush in ZBrush.

Tip
You can change the size of the brush by using the [and] keys. The [key is used to reduce the size of the brush tip while the] key is used to increase the size of the brush tip.

Color

The **Color** palette consists of different buttons and sliders that are used to select solid and gradient colors for filling the canvas or an object.

Document

The options in the **Document** palette are used to open, import, export, resize, and save a ZBrush document. A ZBrush document is a 2D illustration and is saved in the *ZBR* format. If you save a 3D object using the **Document** palette, it will be converted into a 2D illustration and cannot be edited further. You can also import a Photoshop file using the **Document** palette.

Draw

The **Draw** palette consists of different options that can be used to modify the settings of the 3D as well as 2D sculpting brushes. These options enable you to change the size, focal shift, and intensity of the brush stroke. In addition to this, these options help you to determine whether a sculpting brush will raise the surface of an object or push it in.

Edit

The **Edit** palette consists of options that enable you to undo or redo the modifications made in ZBrush. Alternatively, you can press CTRL+Z to undo an operation and CTRL+SHIFT+Z to redo an operation.

File

The options in the **File** palette are used to open and save a ZBrush project. A ZBrush project is a combination of different files. If you save a 3D model using the **Save As** button in the **File** palette, its 2D illustration, Zscript, materials, alphas, and different textures will be saved along with the 3D model. You can also save multiple 3D models in a ZBrush project. The models are saved in the *.ZPR* format.

Note

A ZBrush document is a still 2D image that can be used as an illustration. Therefore, if you save a 3D model as a document, you will not be able to edit it further in the 3D space. To avoid this, you should always save your 3D work using the options available in the ***Tool*** *palette. In ZBrush, a 3D model is referred as ZTool. A ZBrush project contains a ZTool as well as its 2D illustration. A ZBrush project can save multiple ZTools simultaneously. However, it is not recommended to save multiple ZTools in a single ZBrush project as it results in a bigger file size which takes more system resources and slows down the system.*

Layer

The **Layer** palette enables you to work in 16 different layers. Each layer represents a separate canvas. You can merge different layers into a single ZBrush document.

Light

The **Light** palette consists of different types of lights that can be used to light up a scene. Besides this, the **Light** palette also contains different subpalettes that are used to modify light settings in a scene.

Macro

The different options in the **Macro** palette are used to record the series of actions performed in ZBrush. Instead of repeating the same actions again and again, you can save different actions in a macro. The recorded macro can be loaded whenever required.

Marker

The **Marker** palette stores information about the orientation, colors, brush strokes, and position of the ZTools modeled by you. You can redraw a model by using the properties stored in the **Marker** palette.

Material

The **Material** palette consists of a library of different materials that can be applied to an object to give it an appearance of different materials such as glass, silver, water, gold, chrome, and so on. This palette also enables you to modify the settings in the materials as required.

Movie

The options in the **Movie** palette are used to record small movies from the canvas. These movies can be used for illustrating your workflow to other users.

Picker

The options in the **Picker** palette are used to pick information such as color, orientation, depth, and so on from the canvas.

Preferences

The options in the **Preferences** palette are used to customize the ZBrush interface and its behavior. Using these option, you can restore the default interface of ZBrush after making any changes in the interface, load and save hot keys, change the color and layout of the ZBrush interface, and record the ZScript automatically.

Render

The **Render** palette consists of options that are used to render a scene. When you render a scene, the rendered output of the scene will be displayed in the canvas instead of a separate window.

Stencil

The options in the **Stencil** palette are used to create patterns with different shapes. A stencil is used to mark out the areas where strokes will be applied. An alpha can be converted into a stencil by choosing the **Make St** button in the **Transfer** subpalette of the **Alpha** palette, refer to Figure 1-8.

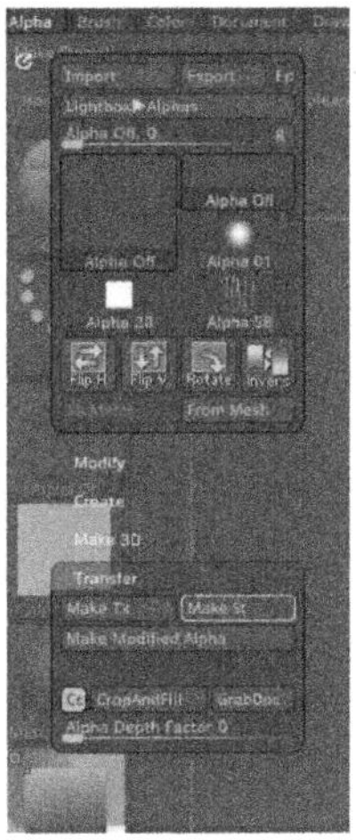

***Figure 1-8** The **Make St** button in the **Transfer** subpalette*

Stroke

The **Stroke** palette consists of options that determine the strength and behavior of brush strokes on the canvas.

Texture

The options in the **Texture** palette help in applying textures to an object. Textures are the 2D images and can be created in ZBrush or can be imported from other sources.

Tool

The **Tool** palette is one of the most important palettes in ZBrush. It consists of various buttons that are used to open, save, import, and export ZTools. In addition to this, it contains a library of inbuilt 3D primitives, ZTools, and 2.5D brushes. When you choose the Current Tool button from the **Tool** palette, a flyout will be displayed, as shown in Figure 1-9.

*Figure 1-9 The flyout displayed on choosing the Current Tool button from the **Tool** palette*

This flyout is divided into three areas namely, **Quick Pick**, **3D Meshes**, and **2.5D Brushes**. The **Quick Pick** area stores a library of 3D primitives and 2.5D brushes that were used recently. The **3D Meshes** area contains a library of default 3D models in ZBrush. It also contains primitive 3D objects such as sphere, cube, rectangle, cone, and so on. You can select any one of these 3D primitives and draw it in the canvas by pressing and holding the left mouse button and then dragging the cursor in the canvas area.

The third area of the **Tool** palette is **2.5D Brushes**. It consists of a library of brushes that are used to add detail to a 2D illustration. The options in the **Tool** palette will be discussed in depth in the later chapters.

Note

If you use a 2.5D brush on a 3D object, the object will be converted into a 2.5D illustration.

Transform

The **Transform** palette consists of different buttons and sliders that are used in modeling, positioning, scaling, rotating, and editing 3D objects.

Zplugin

The options in the **Zplugin** palette are used to access different ZBrush plug-ins that are installed on the computer.

Zscript

ZBrush has an in-built scripting language known as Zscript. ZScripts can be loaded by using the options in the **Zscript** palette.

Trays

Trays are the areas on the ZBrush interface in which you can store different palettes and buttons. The trays are located on the left, right, and bottom of the canvas. By default, the **Tool** palette is docked in the right tray. The left and right trays can be opened or closed by clicking on the arrows adjacent to the left and right shelves. The bottom tray can be opened and closed by double-clicking on the arrow keys below the canvas.

Shelves

Shelves are the areas that contain the most commonly used buttons and controls. The shelves can be classified into top, left, and right shelves. These shelves are discussed next.

Top Shelf

The top shelf is located below the palettes. It consists of different buttons and sliders, as shown in Figure 1-10.

Figure 1-10 *The top shelf*

Home Page

On choosing the **HomePage** button, the **ZBrush: Home Page** window will be displayed. This button is located at the top shelf where the Projection Master used to be in the previous releases.

LightBox

On choosing the **LightBox** button, the **LightBox** browser that contains a library of ZBrush documents, ZTools, ZBrush projects, brushes, alphas, textures, materials, array, and so on will be displayed, refer to Figure 1-11. All ZBrush files can be easily accessed by using the **LightBox** browser, without navigating through different folders on your system. The browser also contains a library of default 3D models which can be modified, as required.

Live Boolean

The **Live Boolean** feature allows you to perform the boolean operations on two or meshes. With the help of this feature, you can create a complex design with ease. To understand the functioning of the boolean operation, choose the Current Tool button from the **Tool** palette; a flyout will be displayed. Choose the **Sphere3D** primitive from this flyout and then drag the cursor on the canvas; a sphere will be created in the canvas. Next, choose the **Edit** button from the top shelf and then choose the **Make PolyMesh3D** button from the **Tool** palette; the sphere will be converted into a polymesh. Now, add another object of your choice from the **Subtool** palette by choosing the **Append** button. Next, choose the **Live Boolean** button from the top shelf and click one of the icons located at the top of the appended primitive in the **Subtool** palette; the boolean operation will be displayed in the viewport. Figure 1-12 shows the boolean operation when the Subtraction icon clicked.

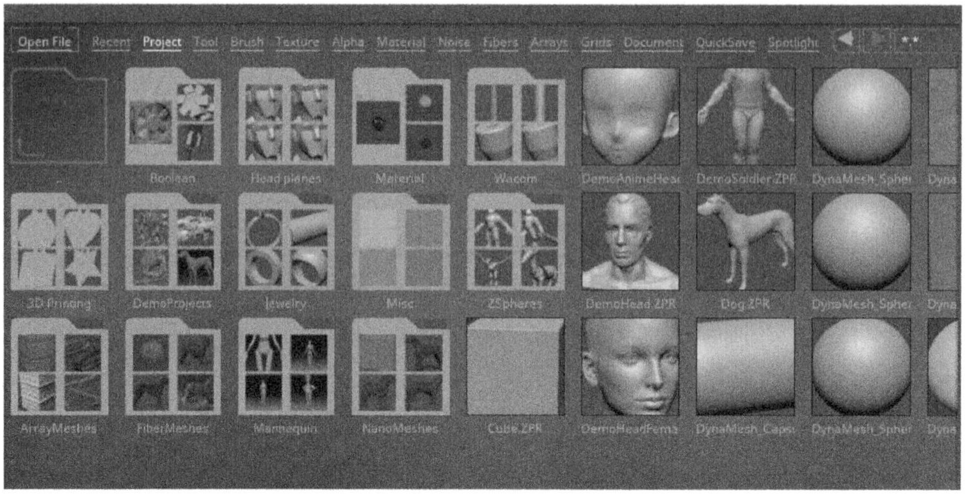

Figure 1-11 *The* ***LightBox*** *browser*

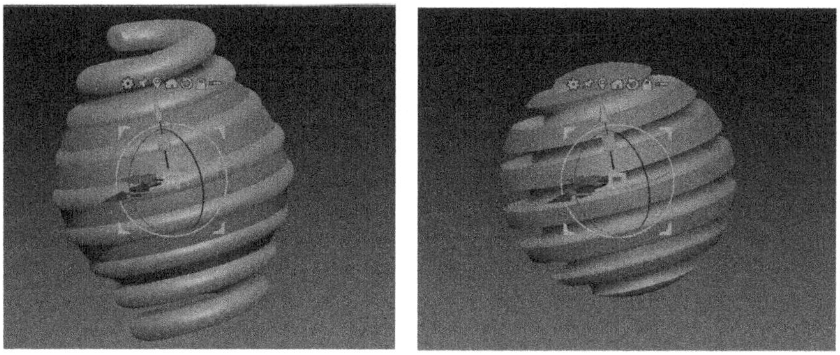

Figure 1-12 *The live boolean operation*

Edit

The **Edit** button is used to enable the sculpting brushes so that you can add details to a 3D model. Whenever a 3D primitive is created in the canvas, you need to choose the **Edit** button from the top shelf. If this button is not chosen, ZBrush will remain in paint mode and dragging the cursor on the canvas will simply place copies of 3D objects in it, refer to Figure 1-13. To make changes in a 3D model, you need to make sure that the **Edit** button is chosen.

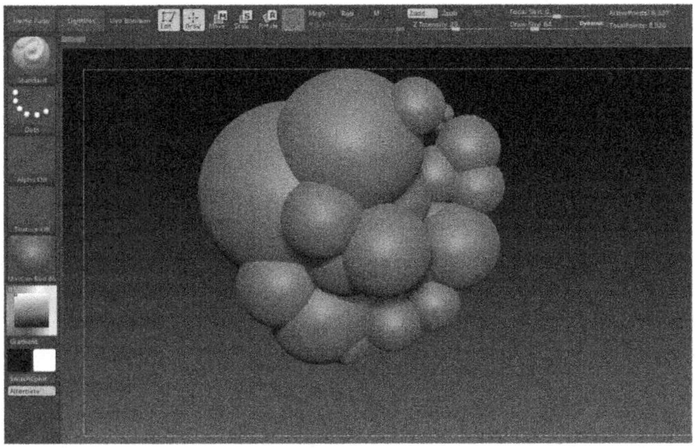

Figure 1-13 *Copies of the 3D objects placed in the canvas*

Tip

You can move an object on the canvas in the edit mode by dragging the cursor on the canvas with the ALT key pressed. To scale an object, press and hold the left mouse button with the ALT key pressed, and then release the ALT key. To rotate an object, press and hold the left mouse button and drag the cursor.

Sculptris Pro Mode

The **Sculptris Pro Mode** button gives you freedom to easily sculpt your model. When you activate this button and start sculpting, the model will be more smoother. The smoothness is determined by the brush size.

Remember Draw Size Per Brush

The **Remember Draw Size Per Brush** button is used to remember the size of brush previously set. By default this button is deactivated. When you activate this button, it allows you to remember the size you set for each sculpting brush excluding the Smooth and Mask brush types.

Remember Dynamic Mode Per Brush

The **Remember Dynamic Mode Per Brush** button is used to control the dynamic size for each brush independently. When this button is deactivated and the Dynamic option located at the right of the Draw size slider is off, the radius of brush changes with the zoom level in the canvas. When the **Remember Dynamic Mode Per Brush** button is activated the radius of the brush remains same even if the zoom level is changed.

Left Shelf

The left shelf consists of different buttons that contain libraries for various brushes, strokes, alphas, textures, materials, and colors, as shown in Figure 1-14. All these libraries can also be accessed through the palettes. However, the left shelf enables easy access to these libraries. The brushes, strokes, alphas, textures, and materials will be discussed in later chapters. The different buttons in the left shelf are discussed next.

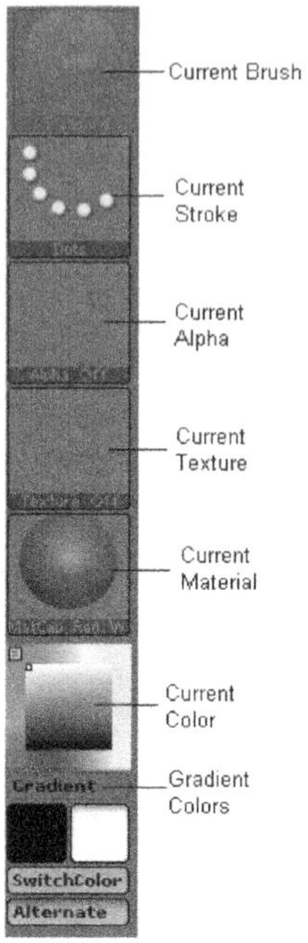

***Figure 1-14** The left shelf*

Current Brush

The Current Brush button is used to invoke a flyout that has different types of sculpting brushes, refer to Figure 1-15. Whenever you select a brush from the flyout, its icon will be displayed on the Current Brush button.

By default, the **Standard** brush icon is displayed on this button. This flyout consists of the **Quick Pick** and **3D Sculpting Brushes** areas. The **Quick Pick** area consists of the recently chosen brushes and the **3D Sculpting Brushes** area consists of all the brushes in ZBrush.

Current Stroke

The Current Stroke button is used to invoke a flyout that consists of different types of strokes, as shown in Figure 1-16. A stroke is a pattern which determines how the painting and sculpting will be done.

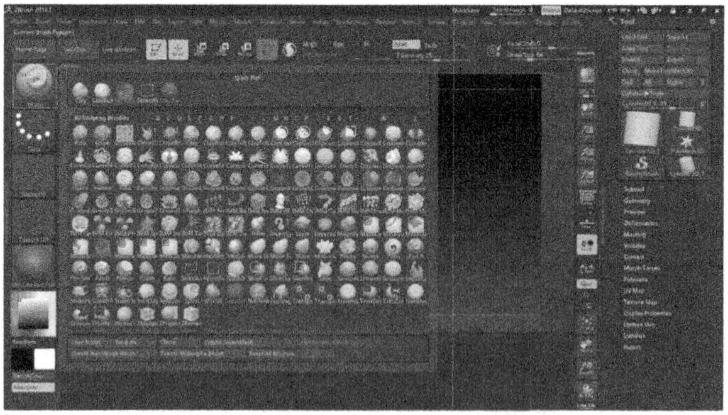

***Figure 1-15** The flyout displayed on choosing the Current Brush button*

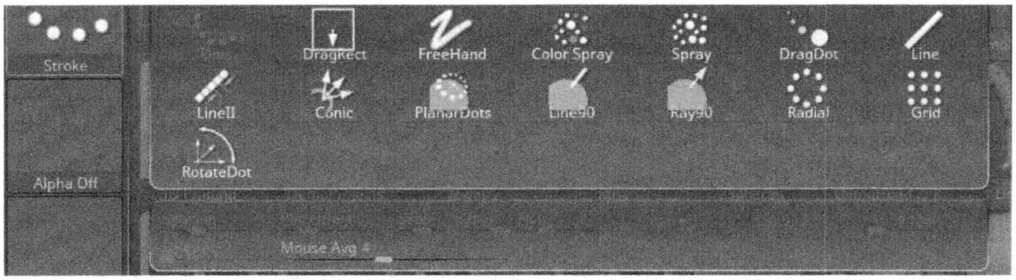

***Figure 1-16** The flyout displayed on choosing the Current Stroke button*

If you select a stroke from the flyout, its icon will be displayed on the Current Stroke button. By default, the **Dots** stroke icon is displayed on this button.

Current Alpha

The Current Alpha button is used to invoke a flyout that consists of different types of alpha images, as shown in Figure 1-17. The different gray scale images in this flyout are known as alphas.

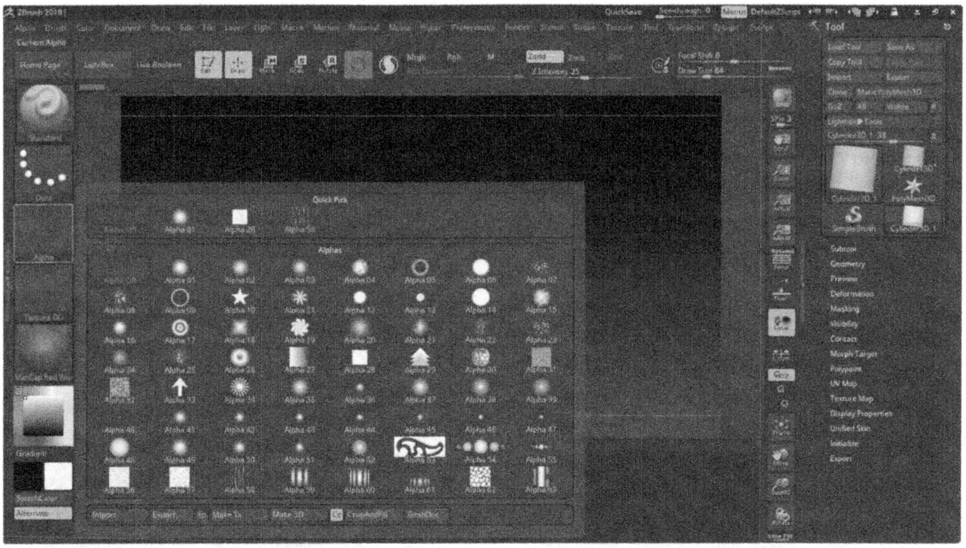

***Figure 1-17** The flyout displayed on choosing the Current Alpha button*

Alphas come in different shapes and determine the shape of brush that will be used in painting and sculpting.

Current Texture

The Current Texture button is used to invoke a flyout that consists of different types of texture images, as shown in Figure 1-18. Textures are the 2D images that can be created in ZBrush or can be imported from other sources.

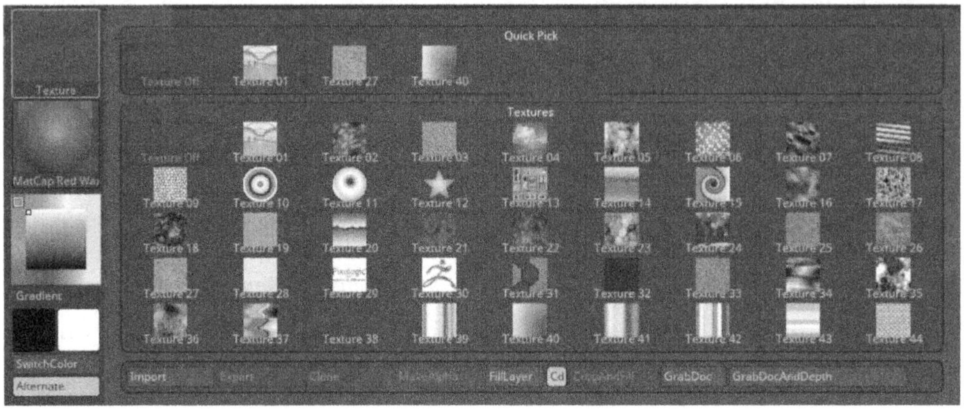

Figure 1-18 The flyout displayed on choosing the Current Texture button

Current Material

The Current Material button is used to invoke a flyout that consists of different types of materials such as glass, silver, water, and gold, as shown in Figure 1-19. These materials can be applied to an object to give it the appearance of the specified material.

Figure 1-19 A flyout displayed on choosing the Current Material button

Current Color

The **Current Color** swatch is located just below the Current Material button. It is used to pick a color from the canvas and fill a 3D object with the selected color. If you hover the cursor on the color picker, the RGB value of the color is displayed beneath the cursor, as shown in Figure 1-20.

Gradient Colors

The **Gradient Colors** button is used to produce a gradient effect with the help of the **Secondary Color** and **Main Color** swatches available below it. The gradient will be produced from the main color to the secondary color. The color swatch below the **Gradient Colors** button represents the secondary and the main color used for gradient filling. The **SwitchColor** button located below this button is used to interchange the primary and secondary colors.

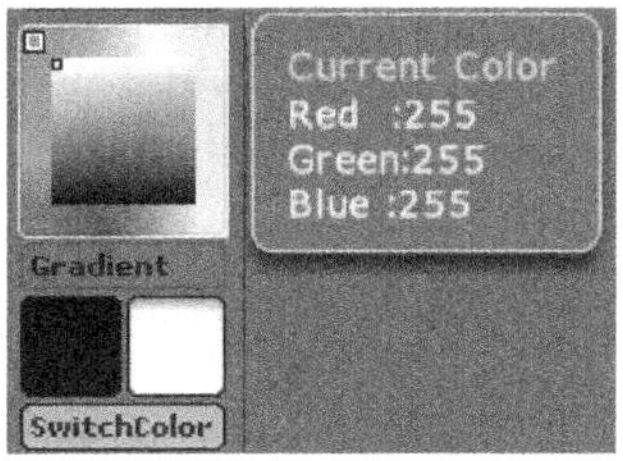

Figure 1-20 *The RGB value of a color displayed*

Right Shelf

The right shelf consists of different buttons and sliders that are used for rendering and navigating through canvas, refer to Figure 1-21.

Figure 1-21 *The right shelf*

BPR

BPR stands for Best Preview Render. This button is used to produce high quality realistic renders on the canvas. It displays details including shadows, depth, transparency, and so on in the rendered scene. This button will render an object only when the **Edit** button is chosen.

SPix

SPix stands for Sub Pixel. This slider controls the quality of antialiasing in the render. Antialiasing is a technique used to smoothen the uneven projections on the edges of an image, thereby producing high quality images. These uneven projections occur because the screen does not have the required resolution to represent a smooth line. The value of the **SPix** slider varies between 0 and 7. If you set the value of the **SPix** slider to 7, the highest quality of antialiasing will be achieved.

Scroll

The **Scroll** button is used to move the canvas. Press and hold the left mouse button on the **Scroll** button and then drag the cursor to scroll the canvas.

OPENING AND SAVING A ZTOOL AND ZBRUSH DOCUMENT

In ZBrush, a 3D model is known as ZTool. In order to save the ZTool, you need to save your 3D model using the **Tool** palette. The model will be saved in the *ZTL* format by choosing the **Save As** button in the **Tool** palette, refer to Figure 1-22. The saved ZTool can be opened later by using the **Load Tool** button in the **Tool** palette.

The ZBrush document only contains the 2D illustration and not the full 3D model. Therefore, if you save a 3D model as a document, only the 2D version of its current view will be saved. This document cannot be edited further in 3D space and can only be used as an illustration. If you want to save a 3D object as an illustration, you must save your 3D model using the **Document** palette by choosing the **Save As** button, refer to Figure 1-23. It will be saved in the *ZBR* format. Once saved, the ZBrush document can be opened by using the **Open** button from the **Document** palette.

You can also save all your ZTools as a project using the **File** palette. To save a project, choose the **File** palette to expand it. In this palette, choose the **Save As** button, refer to Figure 1-24; the **Save Project** dialog box will be displayed, refer to Figure 1-25. In this dialog box, enter the desired name in the **File Name** text box and then choose the **Save** button. It will be saved in the .ZPR format.

*Figure 1-22 The **Save As** button in the **Tool** palette*

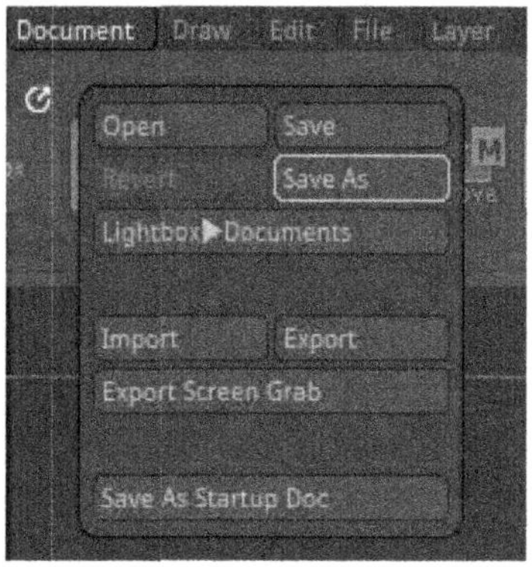

*Figure 1-23 The **Save As** button in the **Document** palette*

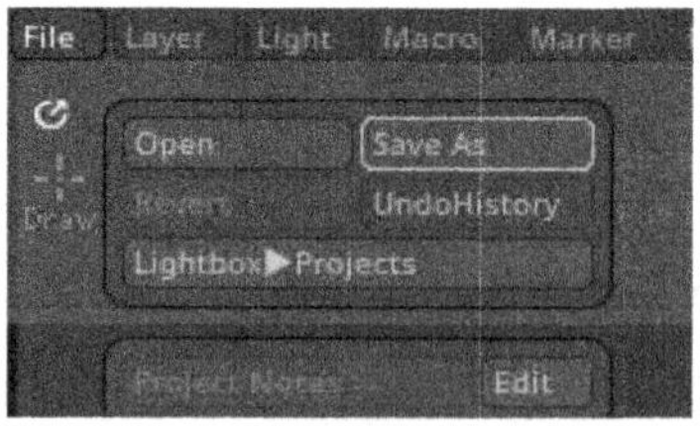

*Figure 1-24 The **Save As** button in the **File** palette*

Tip

On hovering the mouse pointer on different buttons and sliders in ZBrush, the full names of these buttons and sliders will be displayed. If you press and hold the CTRL key while hovering the mouse pointer on the buttons and sliders, a brief description about them will also be displayed.

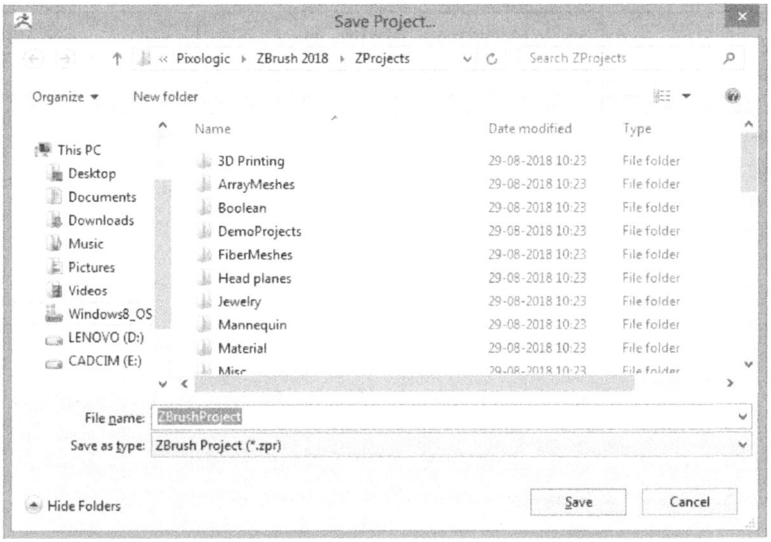

Figure 1-25 *The **Save Project** dialog box*

HOT KEYS

In ZBrush, you can use shortcut keys to invoke the commonly used features. These keys are also referred to as hot keys which help you work efficiently and faster. The most commonly used keys and their functions are listed in Table 1-1.

Table 1-1 *Commonly used keys and their functions*

Key	Function
G	Invokes the **Projection Master** tool
T	Activates the 3D edit mode
Q	Activates the Draw mode
W	Activates the **Move** tool
E	Activates the **Scale** tool
R	Activates the **Rotate** tool

Key	Function
I	Changes the Rgb Intensity
U	Changes the Z Intensity
O	Changes the Focal Shift
S	Adjusts the draw size of the brush
V	Switches colors in the color swatches
0	Displays the actual size of the canvas
P	Activates Perspective Distortion

Chapter 2

Sculpting Brushes

Learning Objectives

After completing this chapter, you will be able to:

- *Understand digital sculpting*
- *Work with different types of sculpting brushes*

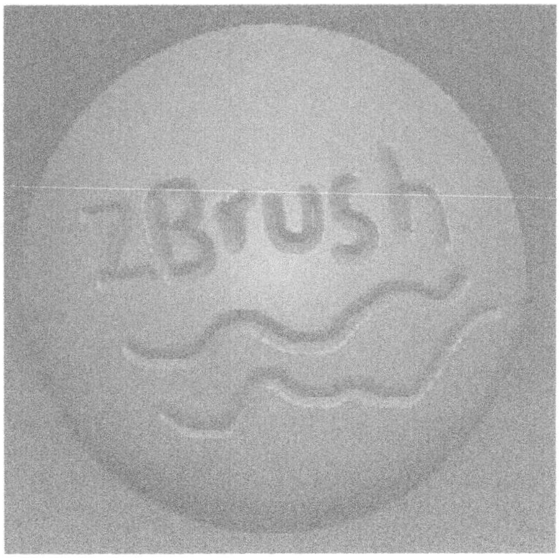

INTRODUCTION

Digital sculpting in ZBrush offers a very efficient and artistic way to create models. It allows you to add subtle details to 3D models which help in achieving photorealistic and hyperrealistic outputs. In this chapter, you will learn about the use of different brushes in sculpting.

Sculpting Brushes

The sculpting brushes are used to modify the shape of a 3D model. These brushes push, pull, smoothen, grab, and pinch a 3D model depending on the type of brush chosen. ZBrush contains an extensive library of brushes. Before using these brushes, you need to create a 3D primitive object or load a model from the LightBox browser. After creating a 3D primitive object or loading a model, you need to choose the **Edit** button in the top shelf. If this button is not chosen, then ZBrush will be in the paint mode and drawing on the canvas will simply create copies of the 3D model in the canvas.

The different brushes in ZBrush can be accessed either from the **Brush** palette located at the top of the interface or using the **Current Brush** button from the left shelf, refer to Figures 2-1 and 2-2. Before using any brush, you can change its settings in the **Brush** palette. In addition to this, you can also modify the brush size and depth by using the settings in the top shelf, refer to Figure 2-3.

Note

*In ZBrush 2018, some of the brushes in the list are shifted to the **Brush** tab of the **LightBox** browser such as **IMM Bricks.ZBP**, **InsertHCylinder.ZBP**, **InsertRing.ZBP**, and so on.*

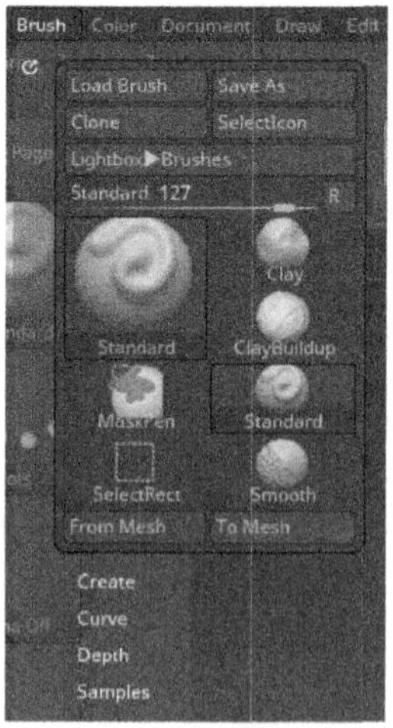

Figure 2-1 *Partial view of the **Brush** palette in the top shelf*

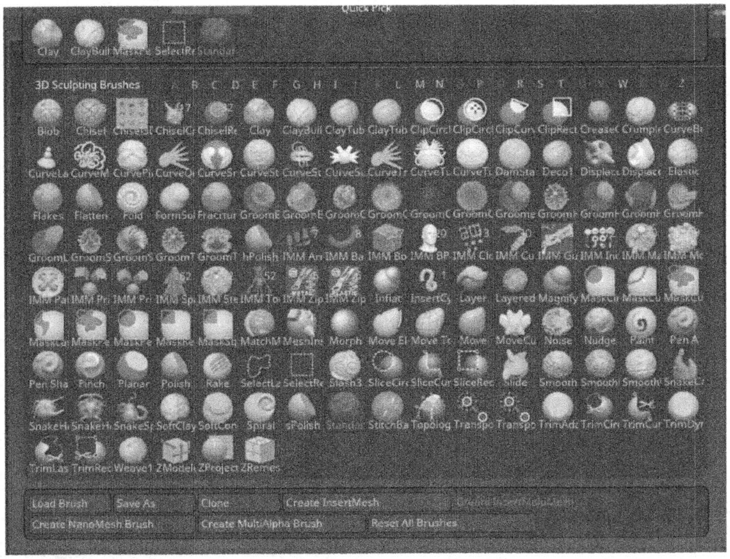

Figure 2-2 The brushes accessed from the left shelf on choosing the ***Current Brush*** *button*

Figure 2-3 *The different brush settings in the top shelf*

After creating a 3D primitive object in the canvas, it should be converted into a polymesh. Polymesh is a collection of polygons or faces that form the surface of a mesh. When you drag a brush on a 3D primitive object, a message box will be displayed, prompting you to convert the 3D primitive object into polymesh, refer to Figure 2-4. You can also import 3D models created in external software applications. These 3D models are imported as polymeshes. The sculpting, texturing, and mapping can only be done on the polymeshes. To convert a 3D primitive object into a polymesh, choose the **Make PolyMesh3D** button from the **Tool** palette, refer to Figure 2-5. After converting a 3D primitive object into polymesh, you can increase the number of polygons so that you can sculpt easily at different resolutions. The various sculpting brushes in ZBrush are discussed next.

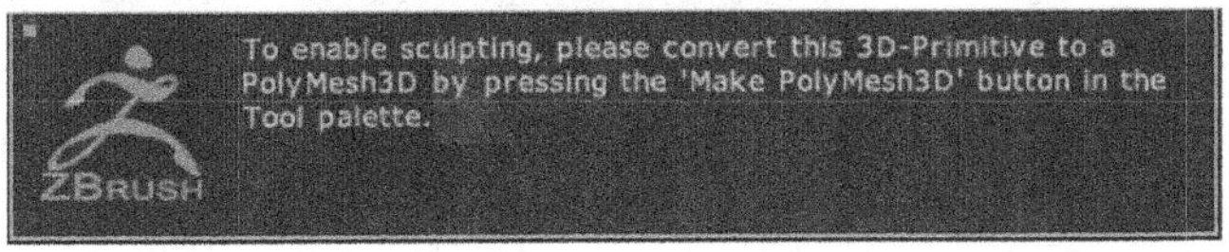

Figure 2-4 *Message box displayed on using a brush on a 3D primitive object*

Note

Before using the brushes on a polymesh, it is recommended to increase the number of polygons in the object. For achieving a greater level of detail on the object, you need to have larger number of polygons. When you place an object on the canvas, it has lower number of polygons. If you *start sculpting on low poly objects, the output will not carry a high level of detail. To increase the number of polygons in the object, expand the **Geometry** subpalette in the **Tool** palette, and then choose the **Divide** button.*

Blob Brush

The **Blob** brush is used to add organic details to an object. This brush is ideal for creating fungus on the surface of an object. To sculpt using this brush, first create a 3D primitive object in the canvas. To do so, choose the **Current Tool** button from the **Tool** palette; a flyout will be displayed, refer to Figure 2-6. Choose the **Sphere3D** primitive from this flyout. Next, press and hold the left mouse button and drag the cursor on the canvas to create a sphere. After creating the sphere, choose the **Edit** button from the top shelf and then choose the **Make PolyMesh3D** button from the **Tool** palette; the primitive object will be converted into a polymesh. Next, expand the **Geometry** subpalette in the **Tool** palette, and then click thrice on the **Divide** button; the sphere will become smoother as more polygons are added, refer to Figure 2-7.

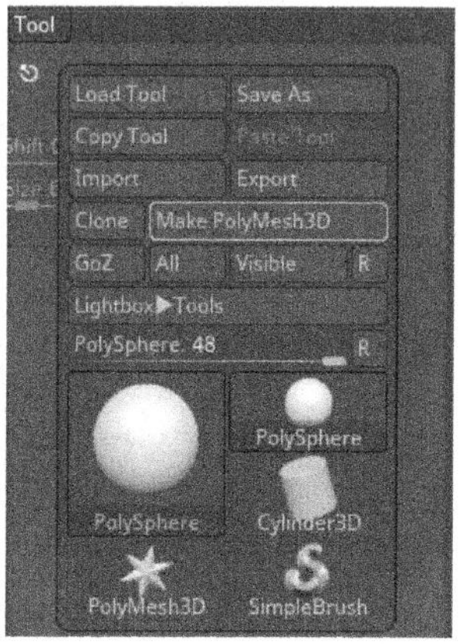

***Figure 2-5** The **Make PolyMesh3D** button in the **Tool** palette*

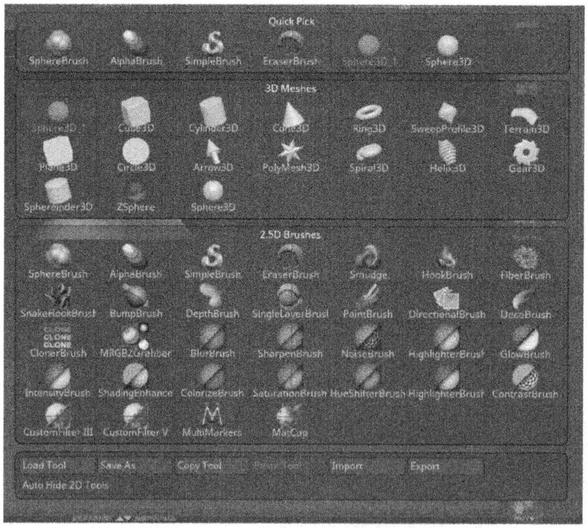

***Figure 2-6** Flyout displayed on choosing the **Current Tool** button*

Choose the Current Brush button from the left shelf; a flyout containing different sculpting brushes will be displayed. Choose the **Blob** brush from this flyout. If required, adjust the size of the brush using the [or] key. Next, press and hold the left mouse button, and drag the cursor on the surface of the sphere; the shape of the sphere will be modified and will become blobby, refer to Figure 2-8.

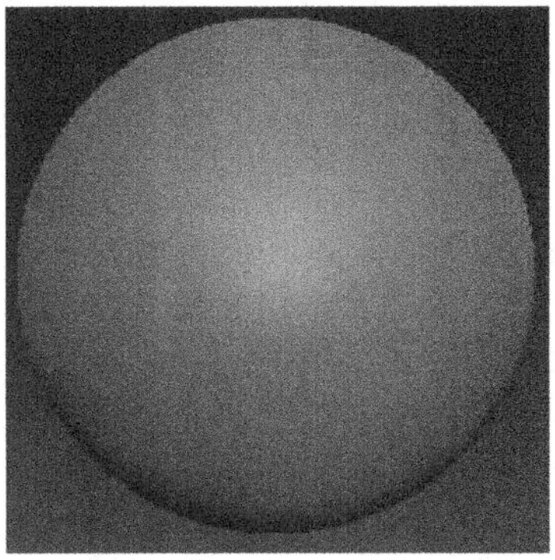

Figure 2-7 The sphere smoothened

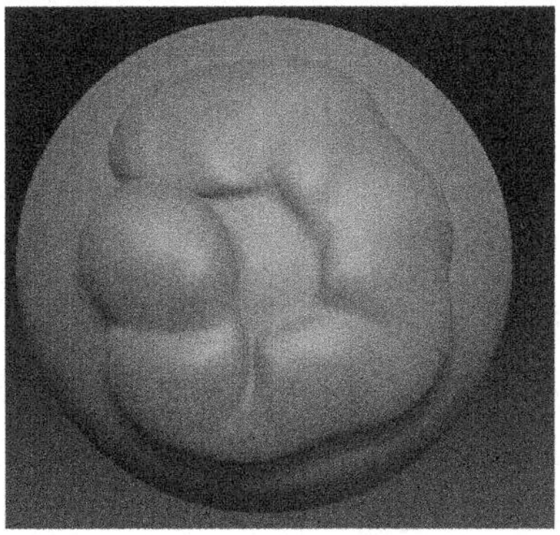

*Figure 2-8 Shape of the sphere modified using the **Blob** brush*

Clay Brushes

There are three types of Clay brushes, namely **Clay**, **ClayBuildup**, and **ClayTubes**, as shown in Figure 2-9. These brushes are discussed next.

Clay Brush

The **Clay** brush is used to add detail and depth to an object by using different alphas, refer to Figure 2-10. On choosing the **Zadd** button, the lump of clay will be created on a sculpture and on an effect will be created as if the sculpture has been scooped on the clay has been taken off, choosing the **Zsub** button.

Figure 2-9 The clay brushes

Figure 2-10 Patterns created using different alphas

ClayBuildup Brush

The **ClayBuildup** brush is used to increase the volume of an object significantly. On choosing this brush, a rectangular alpha gets automatically assigned to it. You can also choose other alpha images from the flyout displayed on choosing the Current Alpha button from the left shelf. To increase the volume of an object using this brush, press and hold the left mouse button and drag the cursor continuously on the surface of the object, refer to Figure 2-11.

ClayTubes Brush
The **ClayTubes** brush is similar to the **ClayBuildup** brush with the only difference that it does not produce displacement in the surface of the object, refer to Figure 2-12. On choosing this brush, a rectangular alpha gets automatically assigned to it. It gives an effect similar to an effect produced when a wooden tool is used on the surface of a wet lump of clay.

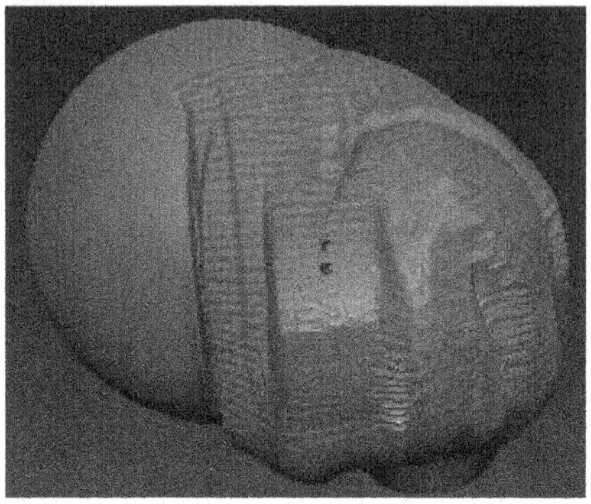

Figure 2-11 *Volume of the sphere increased using the **ClayBuildup** brush*

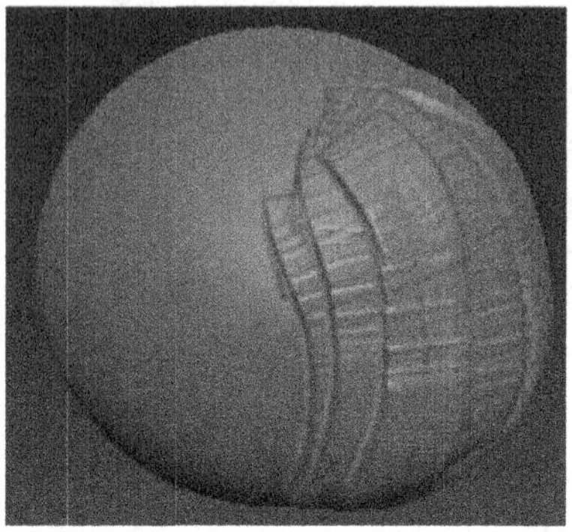

Figure 2-12 *Volume of the sphere increased using the **ClayTubes** brush*

Clip Brushes
The **Clip** brushes are used to slice a particular area of a 3D model, without deleting the polygons in that model. Instead, these brushes squash and squeeze the polygons, without affecting the number of polygons in an object. There are four types of Clip brushes, namely **ClipCircle**, **ClipCircleCenter**, **ClipCurve**, and **ClipRect**, refer to Figure 2-13. These brushes are discussed next.

Figure 2-13 *The **Clip** brushes*

Curve Brushes
There are fifteen types of Curve brushes available in ZBrush, as shown in Figure 2-14. The Curve brushes are used to create different curve patterns on a mesh. These brushes are discussed next.

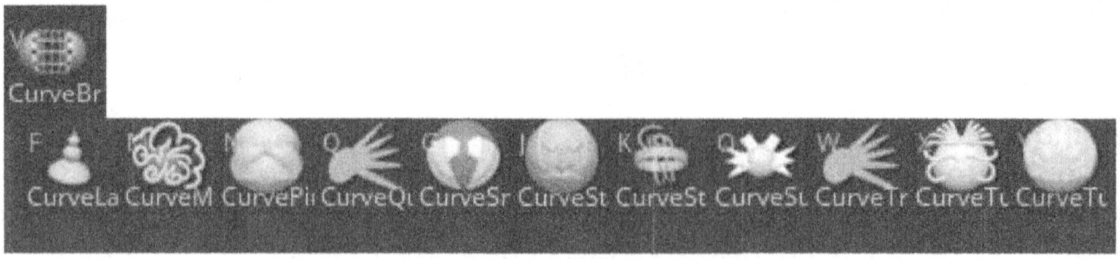

Figure 2-14 *The Curve brushes*

Note

*Before using some of the curve brushes, you need to freeze the subdivision levels of the object. This can be done by choosing the **Del Lower** button from the **Geometry** subpalette of the **Tool** palette.*

CurveBridge Brush

The **CurveBridge** brush is used to create a bridge geometry between two curves. To sculpt using this brush, choose the Current Brush button from the left shelf; a flyout containing different sculpting brushes will be displayed. Choose the **CurveBridge** brush from this flyout. Create a curve on the first sphere and then another curve create on the second sphere; the bridge geometry will be created between two curves on the surface of the spheres, as shown in Figure 2-15.

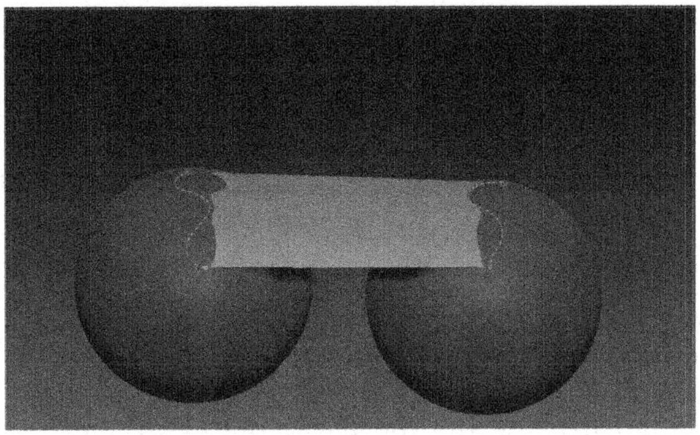

Figure 2-15 *The bridge geometry created between two curves*

CurveLathe Brush

The **CurveLathe** brush is used to create different shapes on the surface of the object. Before using this brush, make sure that the value in the **SDiv** slider is set to 1 or the geometry has not been subdivided. If you use this brush on a mesh which has been subdivided, a message box will be displayed, as shown in Figure 2-16.

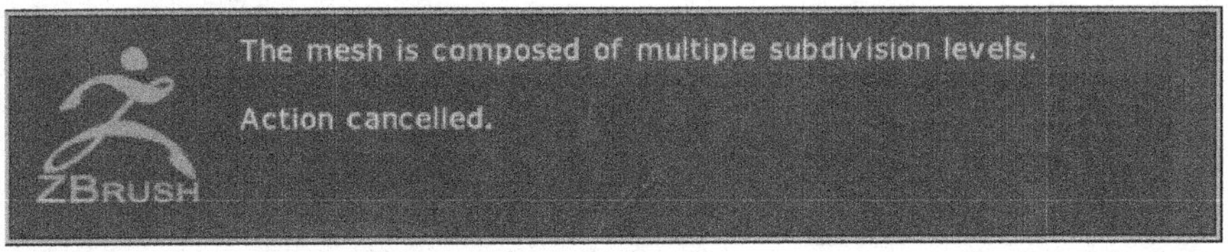

Figure 2-16 *The message box displayed on using the **CurveLathe** brush on subdivided mesh*

To use this brush, create a plane in the canvas and convert it into polymesh. Choose the Current Brush button from the left shelf; a flyout containing different sculpting brushes will be displayed. Choose the **CurveLathe** brush from this flyout. Next, press and hold the left mouse button, and drag the cursor on the surface of the object to create the shape of your choice; a profile curve indicating the shape of the new mesh will be displayed, refer to Figure 2-17. Now, release the left mouse button; a new mesh will be created on the surface, as shown in Figure 2-18.

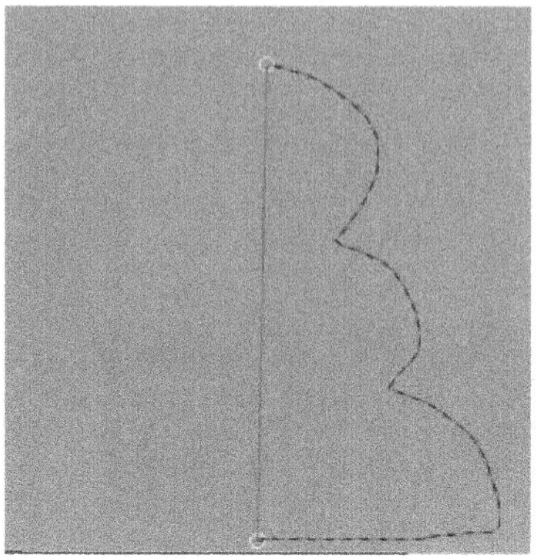

Figure 2-17 *A profile curve created on the surface of plane*

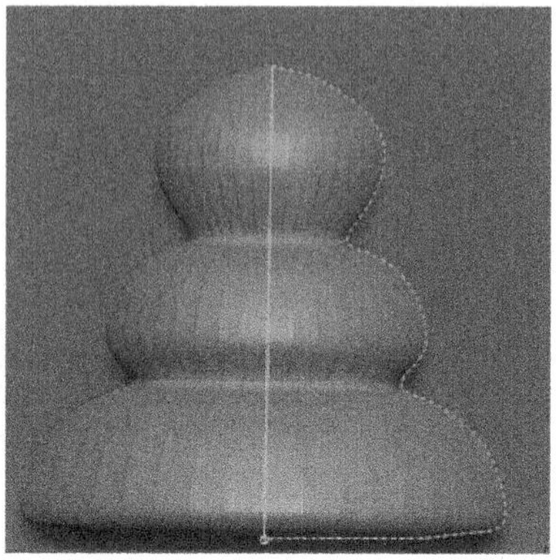

Figure 2-18 *A new mesh created along the profile curve*

You can move, scale, and rotate the newly created mesh by choosing the **Move**, **Scale**, and **Rotate** buttons, respectively from the top shelf. You can also create a duplicate copy of the mesh. To do so, choose the **Move** button from the top shelf. Press and hold the left mouse button and drag the cursor on the surface of the mesh and then release the left mouse button; a line with three circles will be displayed, refer to Figure 2-19. Press and hold the CTRL key and the left mouse button and hover the cursor over the middle circle. Next, drag the cursor toward right; a duplicate copy of the mesh will be created, refer to Figure 2-20.

Figure 2-19 *A line with three circles displayed*

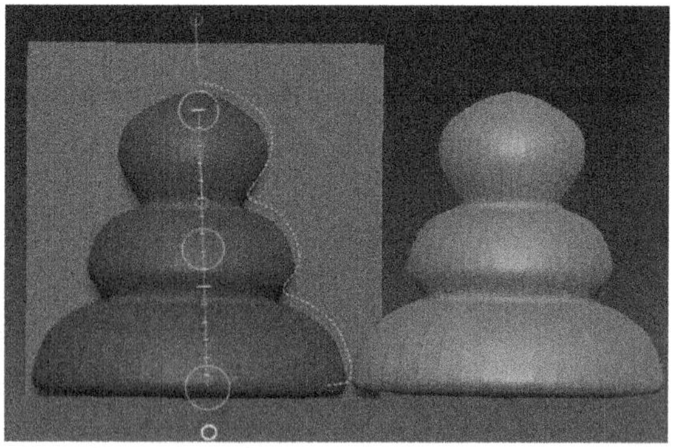

Figure 2-20 *The duplicate copy of the mesh created*

On creating a new mesh using this brush, you will notice that the color of the plane changes to grey and a mask is applied to it. To remove the mask, press and hold the CTRL key, and drag the cursor in the canvas area. After removing the mask, you cannot move, scale, or rotate the new mesh.

Dam_Standard Brush

The **Dam_Standard** brush is used to create very subtle details like wrinkles, clothing seams, machined bevels, and so on. On choosing this brush, the **Zsub** button gets automatically chosen. To use this brush, press and hold the left mouse button, and drag the cursor on the surface of the object; a seam will be created on its surface, refer to Figure 2-21.

Deco1 Brush

The **Deco1** brush is used with different alphas to create detailed decorative designs on an object, refer to Figure 2-22. When this brush is chosen, an alpha gets automatically assigned to it. To sculpt using this brush on a smoothened plane, choose the Current Brush button from the left shelf; a flyout containing different sculpting brushes will be displayed. Choose the **Deco1** brush from this flyout. Adjust the value of the **Draw Size** slider in the top shelf as required. Next, press and hold the left mouse button and drag the cursor slightly at different places on the surface of the plane; a pattern will be created on the surface of the plane.

***Figure 2-21** A seam created in the sphere using the **DamStandard** brush*

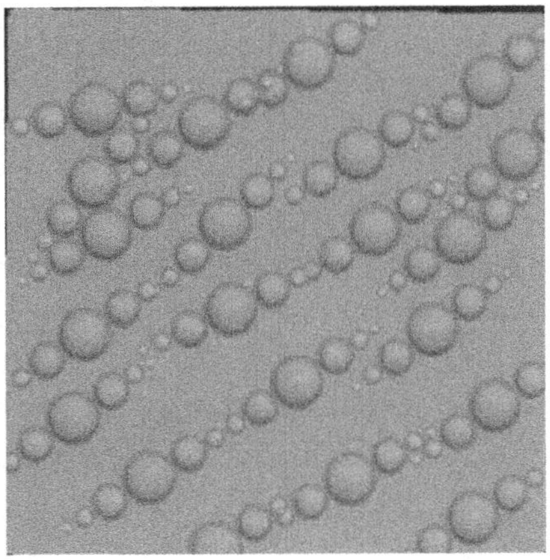

***Figure 2-22** A pattern created on the surface of a plane using the **Deco1** brush*

Displace Brush

The **Displace** brush is used to pull or push the polygons in a direction perpendicular to the surface of an object. To sculpt using this brush, choose the Current Brush button from the left shelf; a flyout containing different sculpting brushes will be displayed. Choose the **Displace** brush from this flyout. Adjust the value of the **Draw Size** slider as required, and set the value of **Z Intensity** slider to **100**. Next, press and hold the left mouse button, and drag the cursor slightly at different places on the surface of the sphere; the polygons will be pulled out perpendicularly, refer to Figure 2-23. To push the polygons in, you need to choose the **Zsub** button from the top shelf and then drag the cursor at different places, refer to Figure 2-24.

DisplaceCurve Brush

The **DisplaceCurve** brush is used to pull or push the polygons in a direction specified by creating a path curve. To sculpt using this brush, adjust the value of the **Draw Size** slider as required, and set the value of the **Z Intensity** slider to **60**. Next, press and hold the left mouse button and drag the cursor on the surface of the object to create the path curve, refer to Figure 2-25. After creating the path curve, hover the cursor on the surface of the path and click on it; the polygons will be pulled out along the path curve, as shown in Figure 2-26.

Figure 2-23 *The polygons pulled out using the **Displace** brush*

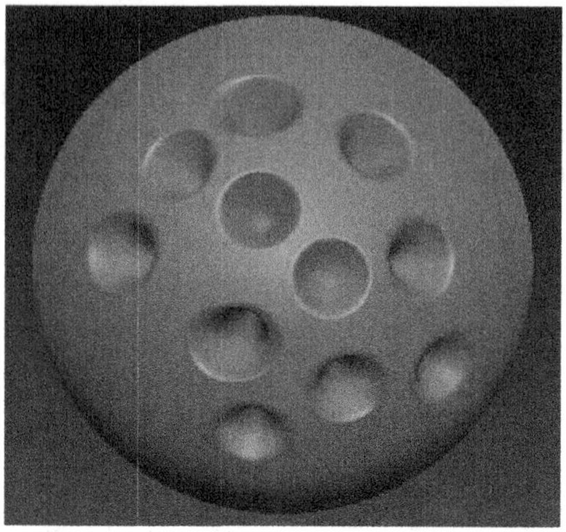

Figure 2-24 *The polygons pushed in using the **Displace** brush*

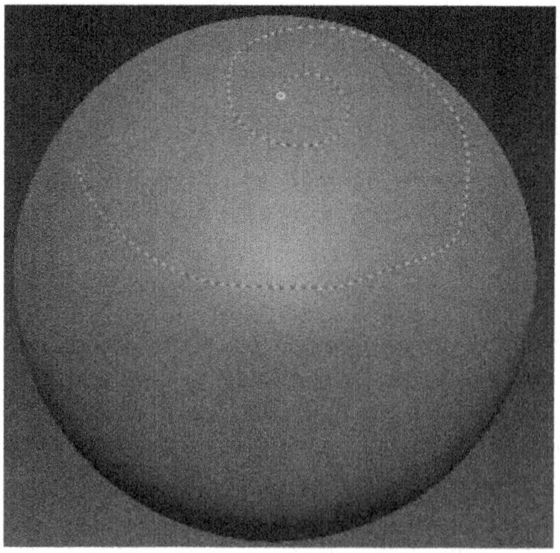

Figure 2-25 *The path curve created on the surface of the sphere*

Figure 2-26 *The polygons pulled out along the path curve*

Elastic Brush

The **Elastic** brush is used to displace the polygons of the surface without changing the original shape of an object, refer to Figure 2-27. In Figure 2-28, you will notice that there is deformation in the surface but the original curves in the sphere are retained.

Flakes Brush

The **Flakes** brush is used to create roughness on the surface of an object. This brush can be used to create rust or flake like effect on a surface, refer to Figure 2-29.

Flatten Brush

The **Flatten** brush is used to flatten the displaced or extruded surface of an object. Figure 2-30 shows a surface on which the **Flakes** brush has been applied and the same surface using the **Flatten** brush on it.

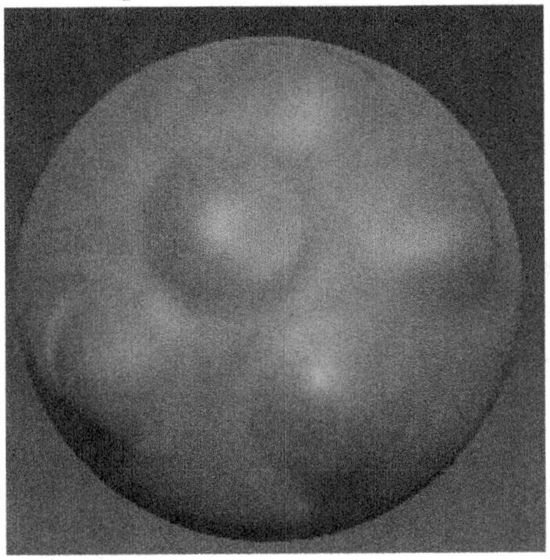

Figure 2-27 *The original shape of the sphere maintained*

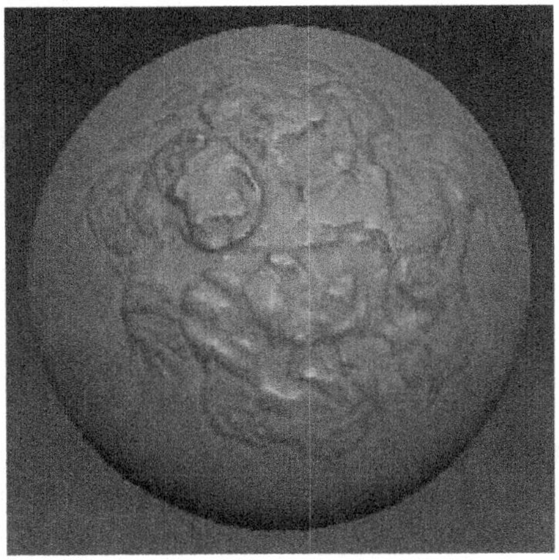

Figure 2-28 *Flakes created on the surface of a sphere*

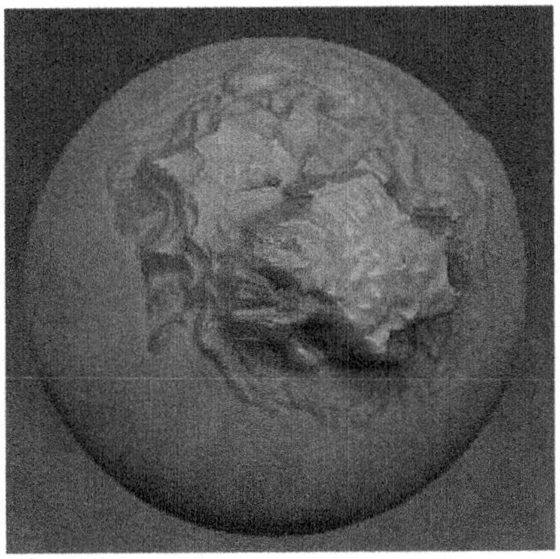

Figure 2-29 *Flakes created on the surface of a sphere using the **Flakes** brush*

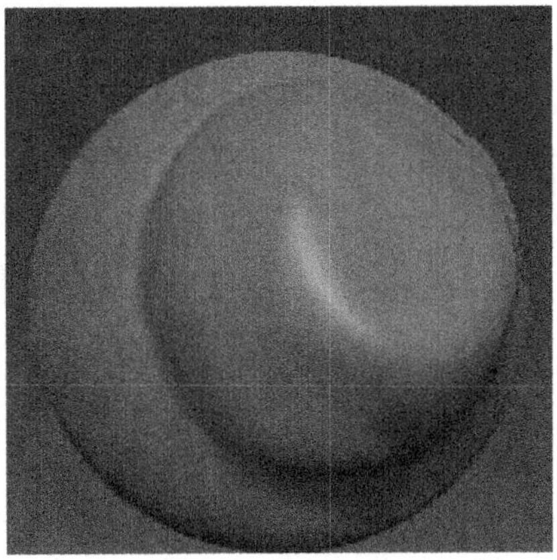

Figure 2-30 *The surface of the sphere flattened using the **Flatten** brush*

Fold Brush

The **Fold** brush is used to fold the polygons inward along the path on which you drag the cursor on the surface of an object, refer to Figure 2-31.

Fracrture Brush

The **Fracrture** brush is used to make the surface of an object disordered and uneven. This brush is best suited for creating rocky surfaces or crystals.

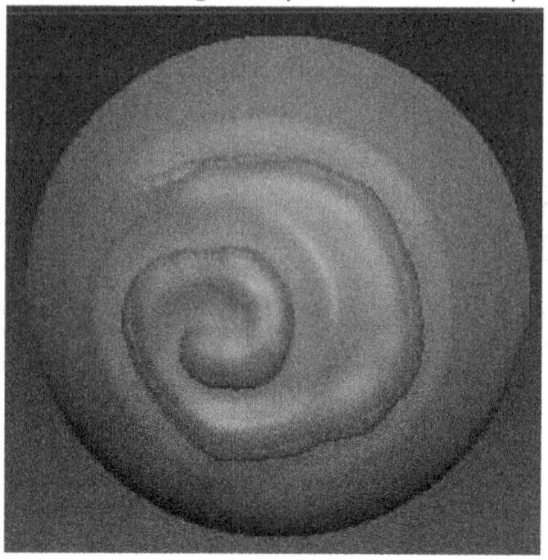

Figure 2-31 *Folds created on the surface of a sphere*

Groom Brushes

The Groom brushes are used to sculpt the **FiberMesh** in ZBrush. The **FiberMesh** is mainly used to create hair on the surface of an object. It is also used to generate different types of fibers, fur, plants, flowers, and so on. The **FiberMesh** option can be accessed from the **Tool** palette.

The Groom brushes have been designed specifically for the **FiberMesh**. These brushes are used to edit fibers in the mesh to create different patterns and hairstyles. There are sixteen types of Groom Brushes, refer to Figure 2-32. These brushes will be discussed in detail in the later chapters.

Figure 2-32 *The Groom brushes*

hPolish Brush

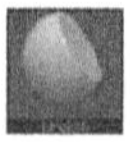

The term **hPolish** stands for hard polish. The **hPolish** brush is used to displace the surface of an object inward, so that the resulting surface has sharp and well defined edges, as shown in Figure 2-33.

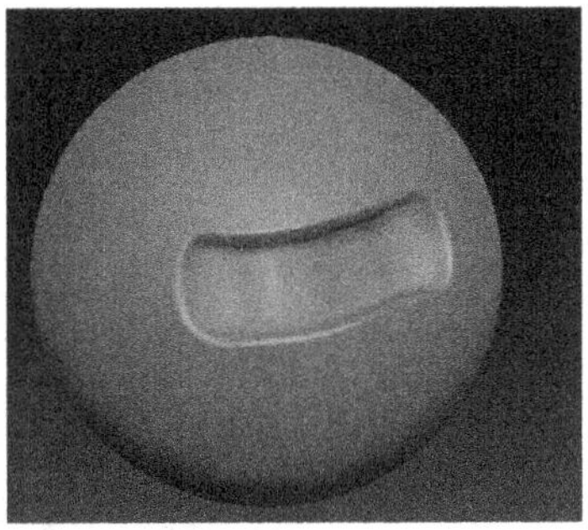

*Figure 2-33 Surface of the sphere modified by using the **hPolish** brush*

IMM Brushes

The term IMM stands for Insert Multi Mesh. The IMM brushes are used to insert different types of objects into an already existing geometry. There are seventeen types of IMM brushes, refer to Figure 2-34. Figure 2-35 shows a sphere with different types of objects inserted into it using IMM brushes.

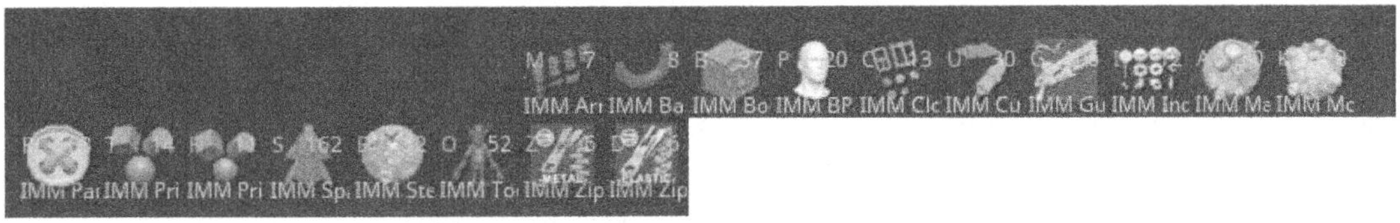

Figure 2-34 The IMM brushes

Note

*Before using some of the **IMM** brushes, you need to freeze the subdivision levels of the object. This can be done by choosing the **Del Lower** button from the **Geometry** subpalette of the **Tool** palette.*

Inflat Brush

The **Inflat** brush is used to expand the surface of an object by a significant amount by pulling the polygons in a direction perpendicular to the surface, refer to Figure 2-36.

Figure 2-35 Different objects inserted into a sphere using IMM brushes

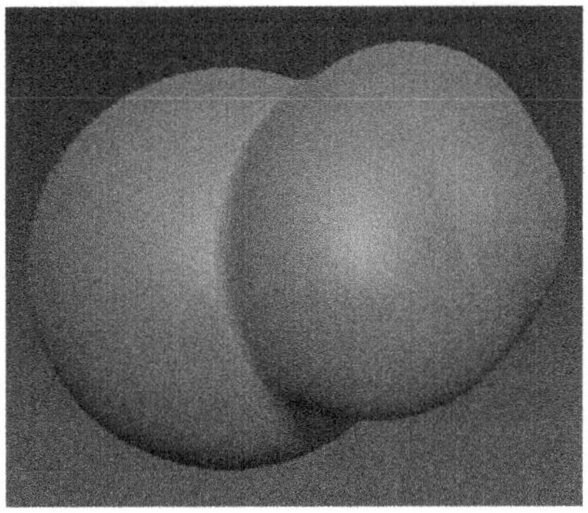

*Figure 2-36 Surface of the sphere expanded by using the **Inflat** brush*

Insert Brushes

The Insert brushes are used to insert different geometrical shapes like cube, sphere, cylinder, and so on, into an already existing geometry. There are eight types of Insert brushes. In ZBrush 2018, two Insert brushes are available in the flyout displayed on choosing the **Current Brush** button and rest of the brushes are in the **Brush** tab of **LightBox** browser, refer to Figures 2-37a and 2-37b. Figure 2-38 shows different shapes inserted into a sphere using these brushes. These brushes work only when the geometry has not been subdivided.

Figure 2-37 a *The Insert brushes*

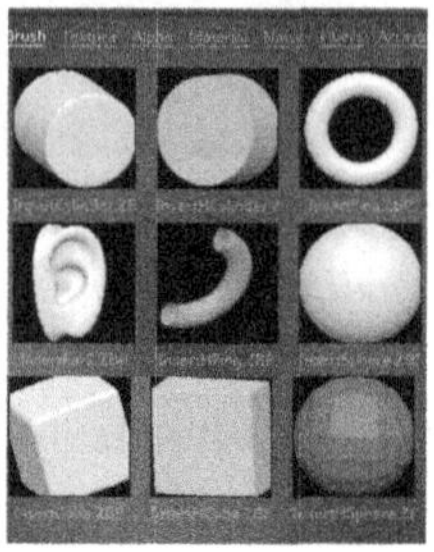

Figure 2-37b *The Insert brushes in Brush tab of **LightBox** browser*

Layer Brush

The **Layer** brush raises or lowers the surface of an object by a fixed amount depending on the value of **Z Intensity**. Using this brush, the depth of displacement remains constant throughout the surface of an object, without any overlapping strokes, refer to Figure 2-39.

Figure 2-38 *Different shapes inserted into a sphere using the insert brushes*

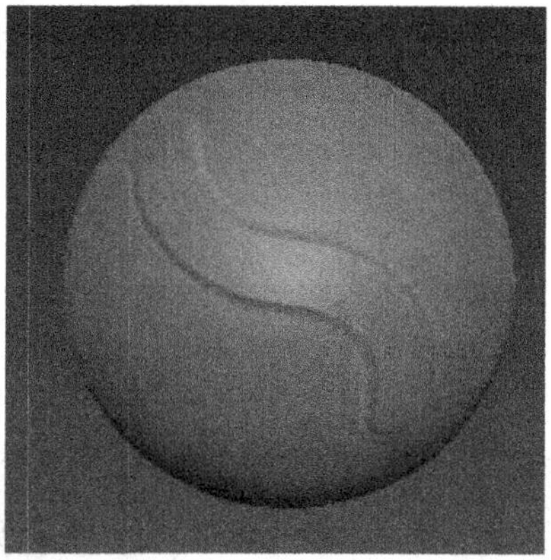

Figure 2-39 *Constant displacement created using the **Layer** brush*

LayeredPattern Brush

The **LayeredPattern** brush is used to create different patterns on the surface of an object, refer to Figure 2-40. On choosing this brush, an alpha is automatically assigned to it. The depth of the pattern created on the surface remains constant throughout the surface.

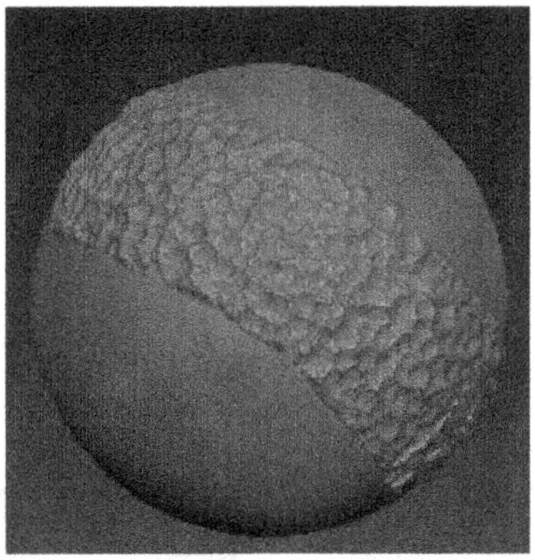

*Figure 2-40 A pattern created using the **LayeredPattern** brush*

Magnify Brush

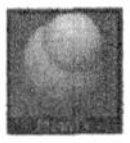

The **Magnify** brush produces significant displacement on the surface of an object. The amount of magnification is determined by the value of the **Z Intensity** slider.

Mask Brushes

The Mask brushes are used to isolate a certain region of an object. If a particular area is masked on a surface, you cannot make any changes in that area. The unmasked area can be modified without affecting the masked area. There are six types of Mask brushes, namely **MaskCircle**, **MaskCurve**, **MaskCurvePen**, **MaskLasso**, **MaskPen**, and **MaskRect**, refer to Figure 2-41.

Figure 2-41 The mask brushes

The **MaskCircle** brush uses a circular stroke to create a mask. The circular area that comes under the mask cannot be sculpted or modified. To use this brush, choose the Current Brush button from the left shelf; a flyout containing different sculpting brushes will be displayed. Choose the **MaskCircle** brush from the flyout; a message box will be displayed prompting you to press the CTRL key to use this brush, refer to Figure 2-42.

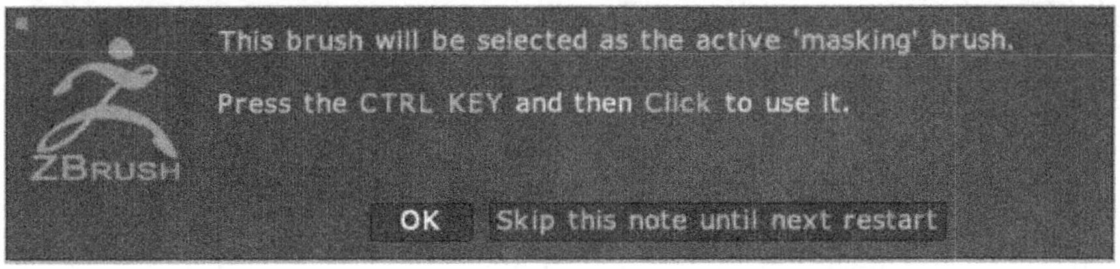

*Figure 2-42 The message box displayed on choosing the **MaskCircle** brush*

Choose the **OK** button to close this message box. Next, press and hold the CTRL key and drag the cursor on the surface of the sphere; a marquee selection will be displayed, refer to Figure 2-43. Next, release the left mouse button; a grey colored circular mask will be created on the surface of the sphere, as shown in Figure 2-44. You can also invert the mask by pressing CTRL+I.

The **MaskCurve** brush uses curves or straight lines to create a mask on the surface of an object. To use this brush, choose the Current Brush button from the left shelf; a flyout containing different sculpting brushes will be displayed. Choose the **MaskCurve** brush from this flyout; a message box will be displayed prompting you to press the CTRL key to use this brush. Choose the **OK** button in this message box.

Next, press and hold the CTRL key and then drag the cursor on the surface of the sphere starting from top to bottom; a straight line with grey highlighted area will be displayed on the surface of the sphere, refer to Figure 2-45. Next, release the left mouse button; a mask will be created on the surface of the sphere, as shown in Figure 2-46.

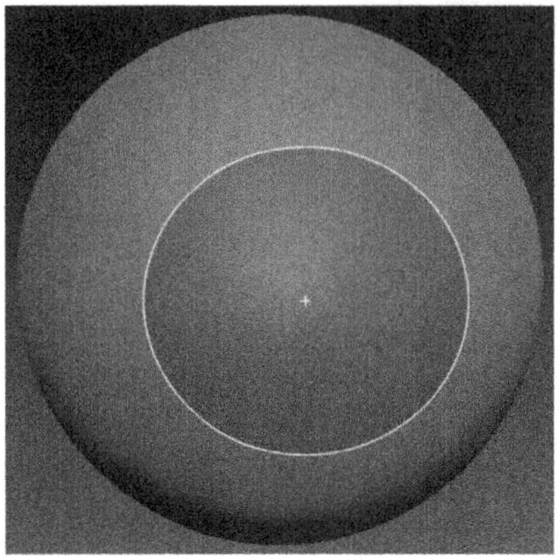

Figure 2-43 *A circular marquee selection displayed*

Figure 2-44 *A circular mask created on the surface of the sphere*

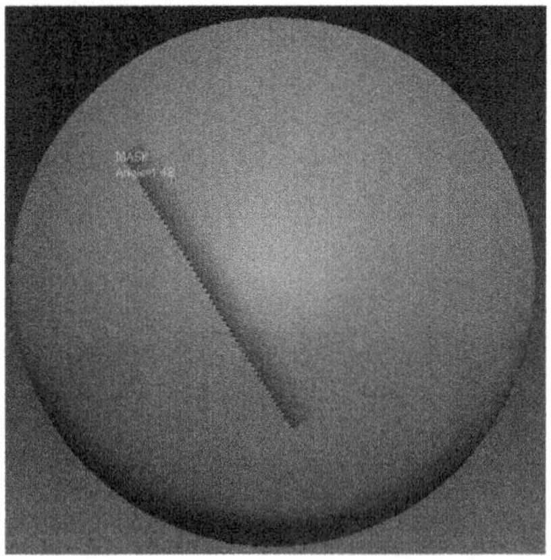

Figure 2-45 *A straight line displayed on the surface of the sphere*

Figure 2-46 *A mask created on the surface of the sphere*

Chapter 3

Introduction to Digital Sculpting

Learning Objectives

After completing this chapter, you will be able to:

* *Understand digital sculpting*
* *Use options in different subpalettes in Tool palette*

INTRODUCTION

In ZBrush, you can create organic as well as inorganic models using the primitive objects. Organic modeling refers to creation of living things such as humans, animals, plants, mythical characters, and insects. Inorganic modeling refers to creation of non-living things such as buildings, cars, and gadgets. ZBrush enables you to create models with complete details using the digital sculpting. In this chapter, you will learn about the process of digital sculpting using different tools and techniques.

TUTORIAL

All the files used in the tutorial can be downloaded from the CADSoft website (*www.cadsofttech.com*). These files are compressed in zip file format and are required to be extracted before using them in the tutorials. The path of the files is as follows: *Textbooks > Animation and Visual Effects > Pixologic ZBrush 2018 > Pixologic ZBrush 2018 for Novices*

Tutorial 1

In this tutorial, you will create a turtle shell using a sphere and different sculpting brushes. The final output of the model is shown in Figure 3-1. **(Expected time: 20 min)**

Figure 3-1 *The turtle shell*

The following steps are required to complete this tutorial:

a. Create the basic shape of the turtle shell.
b. Create divisions in the shell.
c. Add details to the shell.
d. Save the model.

Creating the Basic Shape of the Turtle Shell

In this section, you will create the basic shape of the turtle shell using the *DynaMesh_Sphere_32.ZPR* file located in the **LightBox** browser.

1. Before starting this tutorial, you have to initialize ZBrush to its default state. Choose the **Init ZBrush** button from the **Preferences** palette; ZBrush is initialized to its default state.

2. Double-click on **DynaMesh_Sphere_32.ZPR** in the **Project** tab of the **LightBox** browser, refer to Figure 3-2. The sphere is created on the canvas, as shown in Figure 3-3.

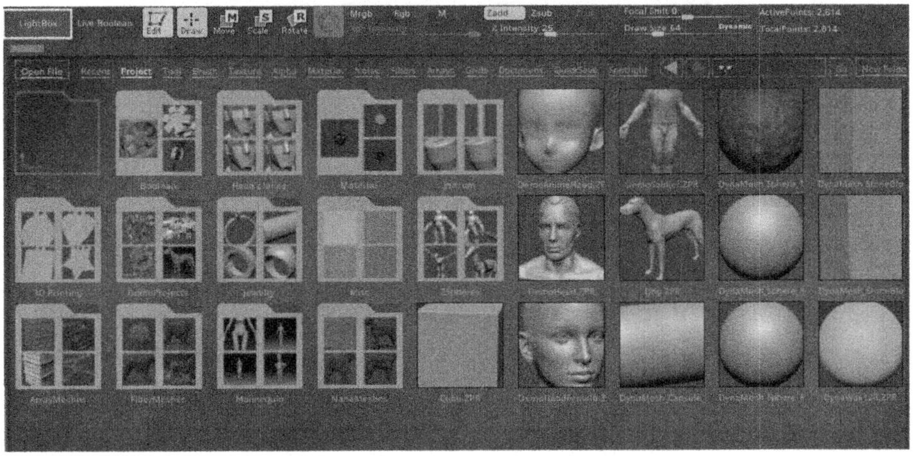

Figure 3-2 *The **DynaMesh_Sphere_32.ZPR** file chosen from the **LightBox** browser*

Figure 3-3 *The **DynaMesh_Sphere_32** model created on the canvas*

In the **DynaMesh_Sphere_32.ZPR** file, the symmetry will be activated in the X-axis by default. As a result, the sculpting done will be mirrored along the X-axis.

Note

*If the **LightBox** browser does not appear on opening the ZBrush window, it can be displayed by choosing the **LightBox** button located in the top shelf, refer to Figure 3-4.*

Figure 3-4 *The **LightBox** button in the top shelf*

3. Choose the Current Brush button from the left shelf; a flyout containing different sculpting brushes is displayed. Choose the **ClipRect** brush from this flyout, as shown in Figure 3-5; a message box is displayed prompting you to press CTRL+SHIFT keys to activate this brush, refer to Figure 3-6. Choose the **OK** button to close this message box.

***Figure 3-5** Choosing the **ClipRect** brush from the flyout*

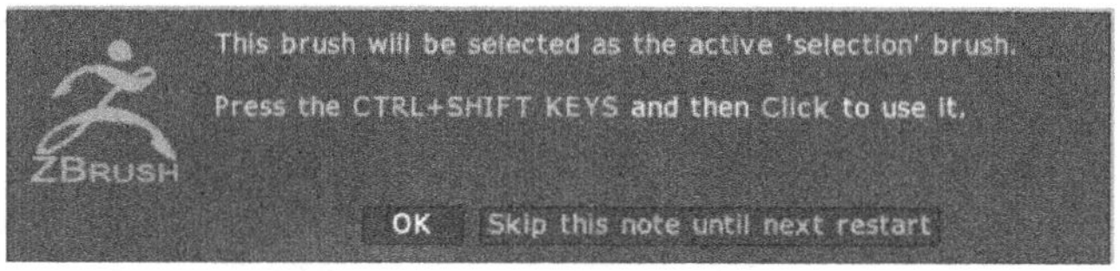

***Figure 3-6** Message box displayed on choosing the **ClipRect** brush*

4. Press CTRL+SHIFT keys and then press and hold the left mouse button. Next, drag the cursor over the sphere; a rectangular marquee selection appears. Select the upper half of the sphere, refer to Figure 3-7. Now, release the left mouse button; the lower half of the sphere is deleted, refer to Figure 3-8.

***Figure 3-7** Upper part of sphere selected using the marquee selection*

***Figure 3-8** Lower half of the sphere deleted*

5. Expand the **Deformation** subpalette in the **Tool** palette, refer to Figure 3-9. You need to increase the size of the sphere along the Z-axis using the **Size** slider.

6. Deactivate the x and y options corresponding to the **Size** slider by choosing the **x** and **y** buttons located on the right side of the slider, refer to Figure 3-10.

7. Set the value of the **Size** slider to **20** by dragging the slider toward right; the sphere is elongated along the Z-axis. Alternatively, you can enter **20** in the edit box displayed on clicking on the slider, refer to Figure 3-11.

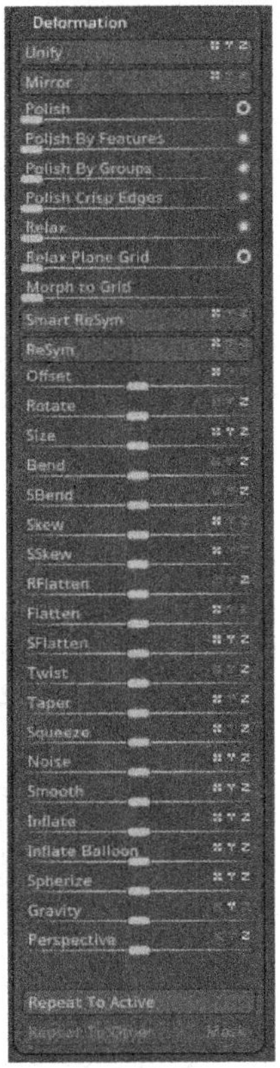

Figure 3-9 The ***Deformation*** *subpalette expanded*

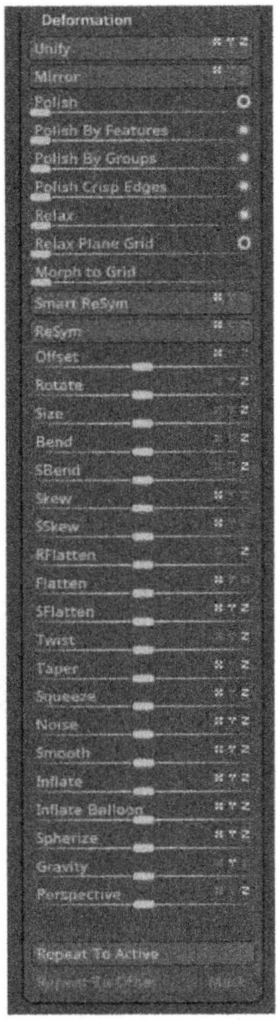

Figure 3-10 *The* ***x*** *and* ***y*** *options deactivated*

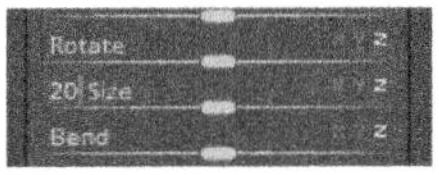

Figure 3-11 *Value entered in the edit box*

8. Expand the **SubTool** subpalette in the **Tool** palette. Next, in this subpalette, choose the **Rename** button; the **Please enter subtool title** window consisting of a text box is displayed. Enter **turtle shell** in this text box and press ENTER; the sphere primitive is renamed as *turtle shell*.

9. To view the elongated *turtle shell*, you need to switch to its top view. To do so, press and hold the left mouse button, and drag the cursor downward in the canvas area; the view is rotated and the top view of the *turtle shell* is displayed, refer to Figure 3-12.

10. Rotate the view of the canvas such that you can view the side view of *turtle shell*. Choose the Current Brush button from the left shelf; a flyout containing different sculpting brushes is displayed. Choose the **Standard** brush from this flyout if not already chosen, as shown in Figure 3-13. Make sure that the **Zadd** button is chosen in the top shelf. Set the value of the **Z Intensity** slider to **20** and of the **Draw Size** slider to **60**, as shown in Figure 3-14.

Figure 3-12 *Top view of the turtle shell*

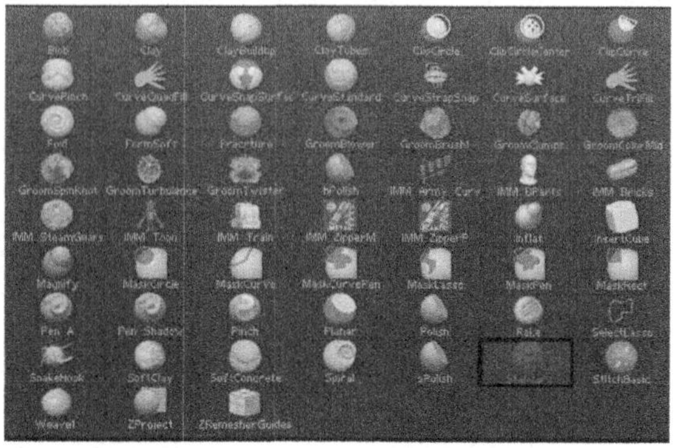

Figure 3-13 *The **Standard** brush chosen from the flyout*

11. Press and hold the left mouse button and drag the cursor on the lower part of the *turtle shell*; the depth is added to the lower part of the shell, as shown in Figure 3-15.

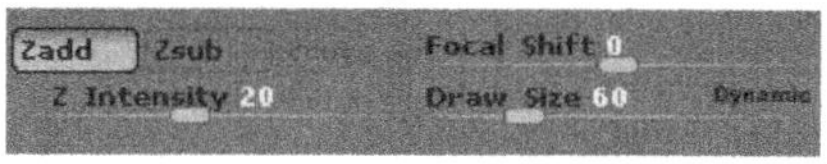

Figure 3-14 *Settings for the **Standard** brush for adding depth*

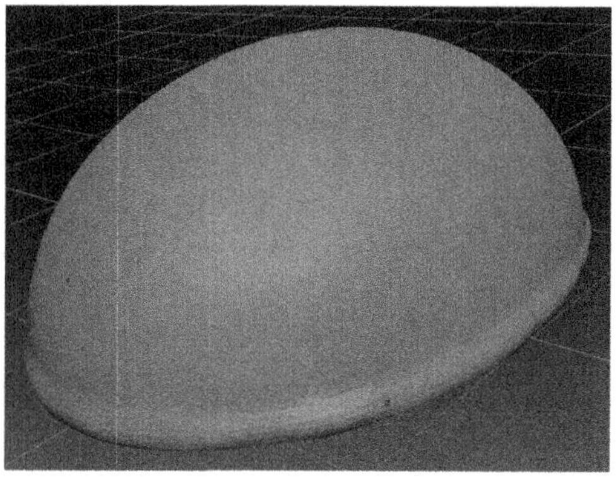

Figure 3-15 *Depth added to the lower part of the turtle shell using the **Standard** brush*

12. Expand the **Geometry** subpalette in the **Tool** palette. In this subpalette, click twice on the **Divide** button; the value of the **SDiv** slider becomes equal to **5**, refer to Figure 3-16, and the *turtle shell* becomes smoother.

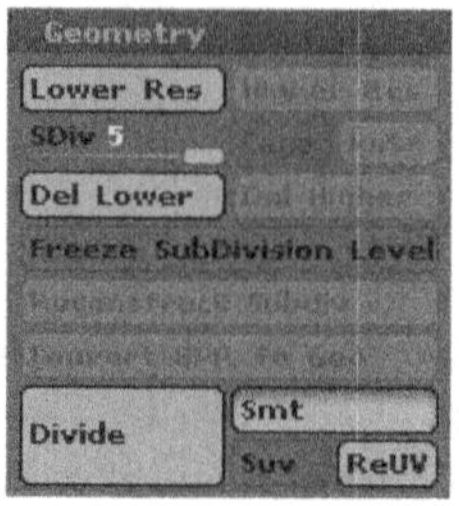

Figure 3-16 *Choosing the **Divide** button from the **Geometry** subpalette*

Creating Patterns on the Turtle Shell

In this section, you will create a pattern on the *turtle shell* using the **Standard** brush.

1. Make sure the **Standard** brush is chosen. Next, choose the **Zsub** button from the top shelf. Set the value of **Z Intensity** slider to **95** and the **Draw Size** slider to **8**, as shown in Figure 3-17.

2. In the **Transform** palette, press the X key to deactivate symmetry in the X-axis.

3. Choose the Current Stroke button from the left shelf; a flyout containing different types of strokes is displayed. Choose the **FreeHand** stroke from this flyout.

4. Press and hold the left mouse button and drag the cursor on the upper part of the *turtle shell* to create a pattern, as shown in Figure 3-18.

Figure 3-17 *Settings for the* ***Standard*** *brush for creating pattern*

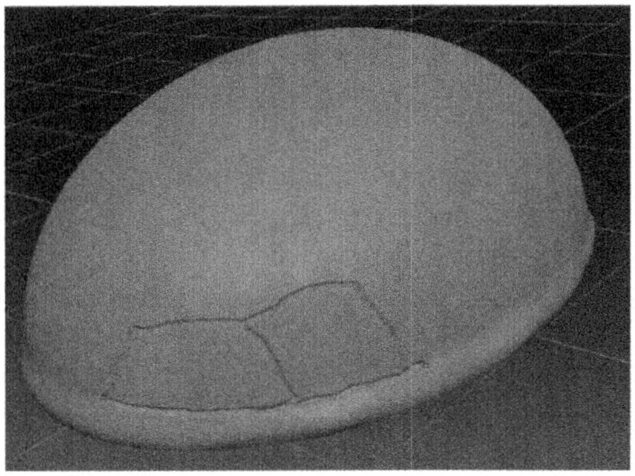

Figure 3-18 *A pattern created on the upper part of the turtle shell using the* ***Standard*** *brush*

5. Continue creating the pattern on the entire upper part of the *turtle shell* using the same settings of the brush, refer to Figure 3-19.

6. Press and hold the left mouse button and drag the cursor to create another pattern on the lower part of the *turtle shell* using the brush settings mentioned above, refer to Figure 3-20.

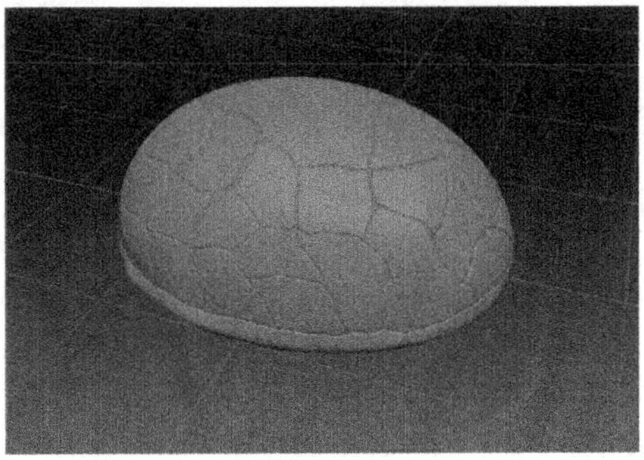

Figure 3-19 *Pattern created on the entire upper part19f the turtle shell*

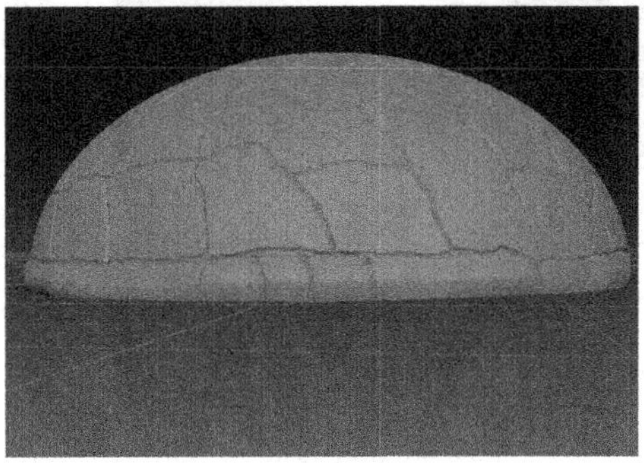

Figure 3-20 *Pattern created on the lower part of the turtle shell using the* ***Standard*** *brush*

7. Continue creating pattern on the lower part of the *turtle shell*, refer to Figure 3-21.

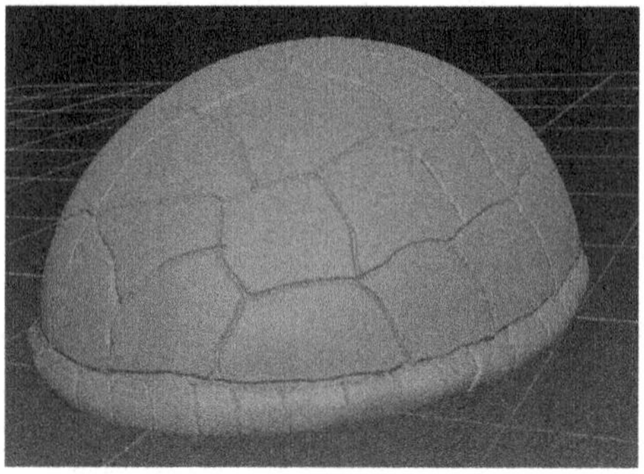

Figure 3-21 *Pattern created on the entire lower part of the turtle shell*

Adding Detail to the Turtle Shell

In this section, you will add detail to the *turtle shell* created in the previous section.

1. Make sure the **Standard** brush is chosen. Choose the **Zadd** button from the top shelf. Set the value of the **Z Intensity** slider to **8** and the **Draw Size** slider to **80**.

   Next, you will add depth to *turtle shell* to create a slightly uneven surface.

2. Press and hold the left mouse button and drag the cursor on the pattern created earlier; a depth is added to the pattern, refer to Figure 3-22.

3. Choose the Current Brush button from the left shelf; a flyout containing different sculpting brushes is displayed. Choose the **Fracture** brush from this flyout, as shown in Figure 3-23.

4. Set the value of the **Z Intensity** slider to **15** and the **Draw Size** slider to **80**.

5. Choose the Current Alpha button from the left shelf; a flyout containing different alpha images is displayed. Choose the **Alpha 07** alpha image from this flyout.

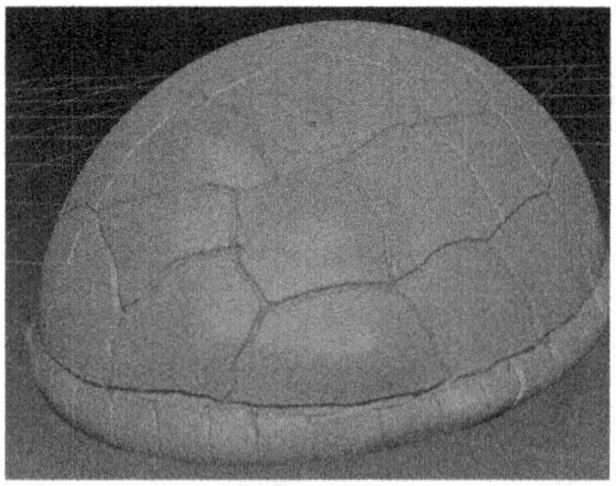

Figure 3-22 *Depth added to the pattern using the* ***Standard*** *brush*

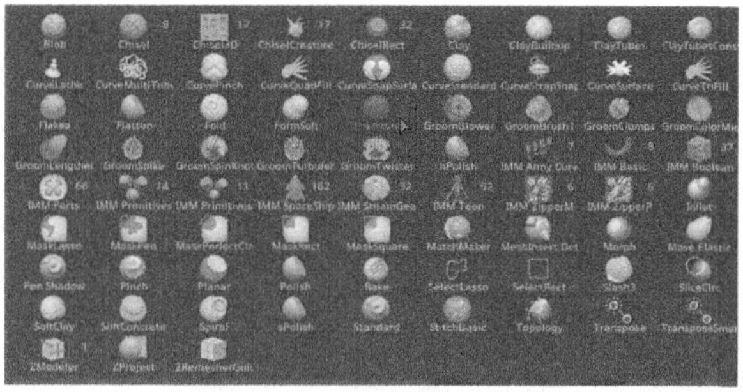

Figure 3-23 *The* ***Fracture*** *brush chosen*

6. Press and hold the left mouse button and drag the cursor on the surface of the *turtle shell*; the final output is created, as shown in Figure 3-24.

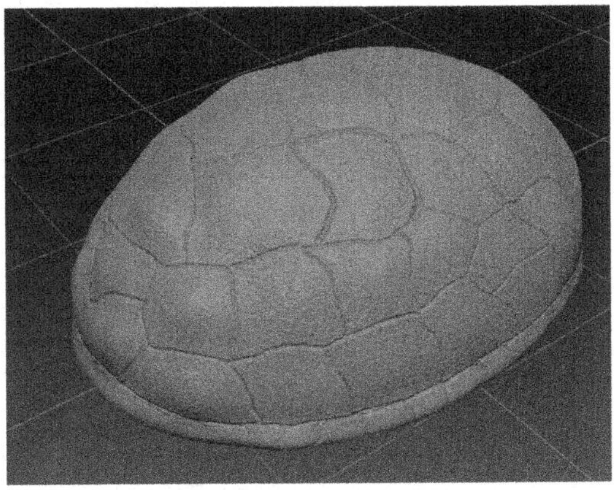

Figure 3-24 *The final output*

Saving the Model

In this section, you will save the file using the steps given next.

1. Choose the **Save As** button from the **Tool** palette; the **Save ZTool** dialog box is displayed. In this dialog box, browse to the location *\Documents\ZBrushprojects\c03*.

2. Enter **c03tut1** in the **File name** edit box and then choose the **Save** button.

EXERCISES

The output of the models used in the following exercises can be accessed by downloading the *c03_ZBrush_2018_exr.zip* file from *www.cadsofttech.com*. The path of the file is as follows: *Textbooks > Animation and Visual Effects > ZBrush > Pixologic ZBrush 2018 for Novices*.

Exercise 1

Create the model of a table lamp using radial symmetry, as shown in Figure 3-25.

(Expected time: 15 min)

Figure 3-134 *Model of a table lamp*

Exercise 2

Create the model of a photo frame using the **Projection Master** dialog box, as shown in Figure 3-26. **(Expected time: 30 min)**

Figure 3-26 *Model of a photo frame*

Chapter 4

SubTools and FiberMesh

Learning Objectives

After completing this chapter, you will be able to:

- *Work with the SubTool subpalette*
- *Work with the FiberMesh subpalette*
- *Understand different settings of FiberMesh*

INTRODUCTION TO SUBTOOLS IN ZBrush

In ZBrush, subtools are sub components of an object. You can work individually on different components by selecting a subtool from the **SubTool** subpalette. For instance, to model a human head, you can model its face, eyes, and teeth separately. After modeling, you can work on each one of the subtools separately without disturbing the other subtools. Subtools play an important role in modeling complex objects.

TUTORIALS

All the files used in the tutorial can be downloaded from the CADSoft website (*www.cadsofttech.com*). These files are compressed in zip file format and are required to be extracted before using them in the tutorials. The path of the files is as follows: *Textbooks > Animation and Visual Effects > Pixologic ZBrush 2018 > Pixologic ZBrush 2018 for Novices*

Tutorial 1

In this tutorial, you will create an ice cream cone and sculpt it using different brushes. The final output of the ice cream model is shown in Figure 4-1. **(Expected time: 20 min)**

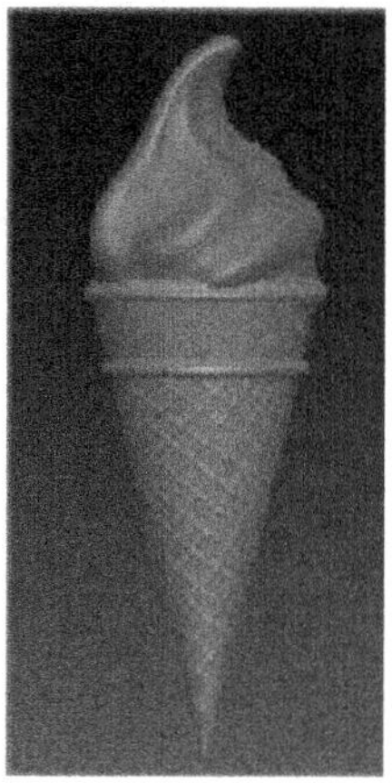

Figure 4-1 *The ice cream cone*

The following steps are required to complete this tutorial:

a. Create the cone.
b. Create the ice cream scoop.
c. Save the model.

Creating the Cone

In this section, you will create an ice cream cone using the **Cone 3D** primitive and sculpt it using different brushes.

1. Choose the **Init ZBrush** button from the **Preferences** palette; the message box is displayed . Choose the **Yes** button from the message box; ZBrush is initialized to its default state. To close the *LightBox* browser, choose the **LightBox** button.

2. Choose the Current Tool button from the **Tool** palette; a flyout is displayed. Choose the **Cone3D** primitive from this flyout.

3. Press and hold the left mouse button, and then drag the cursor in the canvas; a cone is created in the canvas, as shown in Figure 4-2.

4. Choose the **Edit** button from the top shelf.

5. Press and hold the left mouse button and drag the cursor downward in such a way that the cone rotates and stands vertically. While dragging the cursor, press the SHIFT key; the cone is snapped at right angle to the canvas, refer to Figure 4-3.

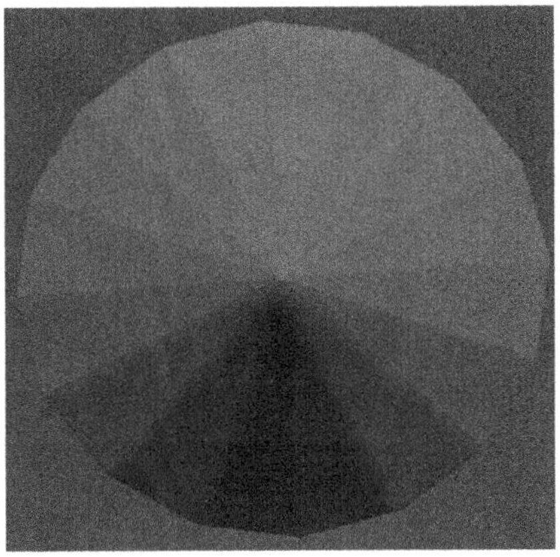

***Figure 4-2** The cone created in the canvas*

***Figure 4-3** The cone snapped to the canvas*

6. Expand the **Initialize** subpalette in the **Tool** palette. In this subpalette, set the values of both the **X Size** and **Y Size** sliders to **50**, refer to Figure 4-4; the radius of the cone decreases.

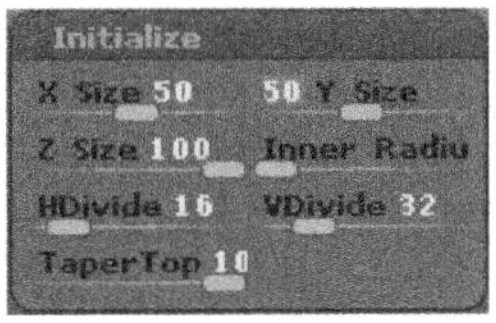

***Figure 4-4** The values in **X Size** and **Y Size** sliders set*

7. Choose the **Make PolyMesh3D** button from the **Tool** palette; the cone is converted into a polymesh.

   After converting the cone into a polymesh, you need to subdivide the geometry so that the number of polygons in the cone increases to make it smoother.

8. Set the value of the **SDiv** slider in the **Geometry** subpalette to **5** by clicking on the **Divide** button four times; the cone becomes smoother, refer to Figure 4-5.

9. In the **SubTool** subpalette, choose the **Rename** button; the **Please enter subtool title** window consisting of a text box is displayed. Enter **ice cream cone** in this text box and press ENTER; the **Cone3D** primitive is renamed as *ice cream cone*.

10. In the **Transform** palette, choose the **Activate Symmetry** button. Next, choose the **(R)** button; the **RadialCount** slider is activated. By default, the **>X<** button is chosen. To deactivate this button, choose it again. Next, choose the **>Z<** button; the radial symmetry in the Z-axis is activated. Click on the **RadialCount** slider to display the **RadialCount** edit box. Next, enter the value **60** in the **RadialCount** edit box, refer to Figure 4-6.

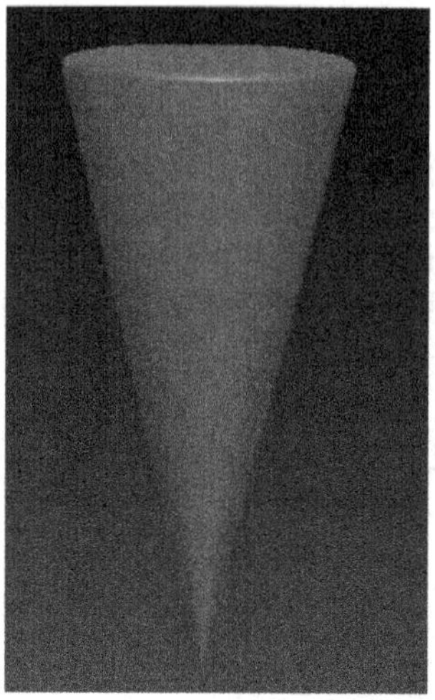

Figure 4-5 *The ice cream cone smoothened*

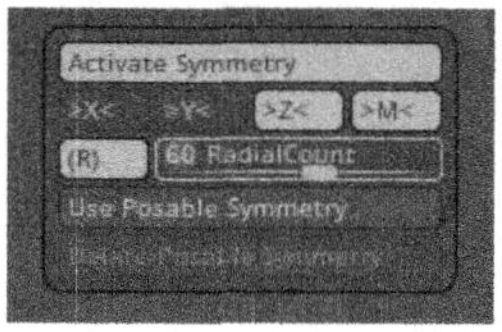

Figure 4-6 *The value of the* ***RadialCount*** *slider set to* ***60***

11. Choose the Current Brush button from the left shelf; a flyout containing different sculpting brushes is displayed. Choose the **Standard** brush from this flyout.

12. Set the value of the **Draw Size** slider to **30**. Next, press and hold the left mouse button, and create the patterns at the top of the *ice cream cone*, as shown in Figure 4-7.

13. Choose the Current Brush button from the left shelf; a flyout containing different sculpting brushes is displayed. Choose the **Layer** brush from this flyout. In the **Transform** palette, choose the **(R)** and **>Z<** buttons; the radial symmetry in the Z-axis is deactivated.

14. Expand the Stroke palette. In this palette, expand the **Lazy Mouse** area. Next, choose the **LazyMouse** button in this area, if not already chosen, as shown in Figure 4-8.

    The **LazyMouse** button will enable you to draw the patterns very conveniently by dragging the red colored line that appears on choosing the **LazyMouse** button.

15. Set the value of the **Z Intensity** and **Draw Size** sliders to **40** and **12**, respectively. Next, press and hold the left mouse button and drag the cursor to create a criss-cross pattern on the entire surface of *ice cream cone*, refer to Figure 4-9.

Tip

In order to get straight strokes using the ***LazyMouse*** *button, you need to press the SHIFT key while dragging the red colored line on the surface of an object.*

Figure 4-7 Patterns created at the top of the ice cream cone

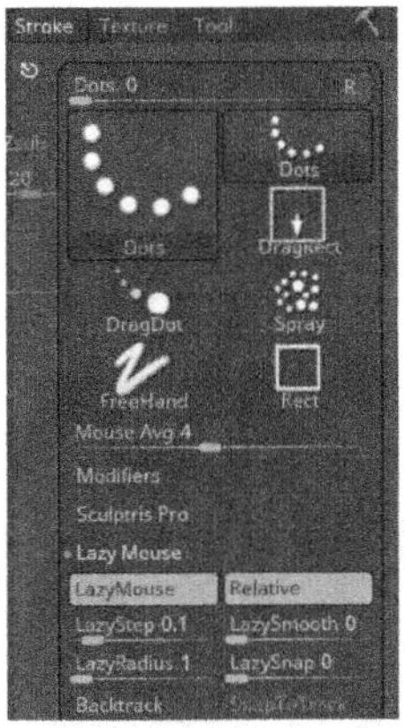

*Figure 4-8 The **LazyMouse** button chosen from the **Stroke** palette*

Creating the Ice Cream Scoop

In this section, you will create the ice cream scoop using the sphere and different sculpting brushes.

1. Expand the **SubTool** subpalette in the **Tool** palette. In this subpalette, choose the **Append** button, as shown in Figure 4-10; a flyout is displayed. Choose the **Sphere3D** primitive from this flyout; a sphere is created in the canvas and it overlaps *ice cream cone*, as shown Figure 4-11. The thumbnail for the sphere is displayed in the **SubTool** list. Select the thumbnail of the sphere in the list and then rename it to **scoop** using the **Rename** button.

2. Choose the **Gizmo 3D Y** button from the top shelf to deactivate it. Next, choose the **Move** button from the top shelf; an action line is displayed. Next, press and hold the left mouse button and drag the cursor on the surface of the sphere; the size of the action line increases.

3. Hover the cursor at the centre of the middle circle of the action line, refer to Figure 4-12. Next, press and hold the left mouse button and drag the cursor upward to move the sphere at the top of *ice cream cone*.

4. In the **Geometry** subpalette of the **Tool** palette, set the value of the **SDiv** slider to **5** by clicking on the **Divide** button four times; the sphere becomes smoother.

Figure 4-9 *The crisscross pattern created on ice cream cone*

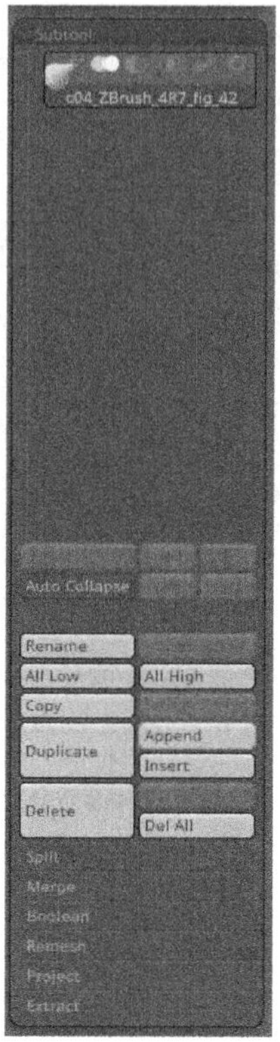

Figure 4-10 *The* ***Append*** *button chosen in the* ***SubTool*** *subpalette*

Figure 4-11 *The sphere overlapping ice cream cone*

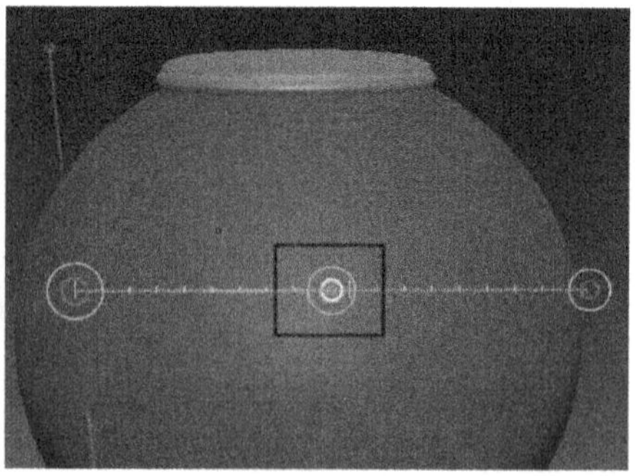

Figure 4-12 *The action line*

5. Expand the **Deformation** subpalette in the **Tool** palette. In this subpalette, move the **Size** slider toward left so that the size of *scoop* decreases and fits into the radius of the cone, as shown in Figure 4-13.

6. Choose the **Draw** button from the top shelf and then choose the Current Brush button from the left shelf; a flyout containing different sculpting brushes is displayed. Choose the **Move** brush from this flyout.

7. Set the value of the **Draw Size** slider in the top shelf to **250**. Next, press and hold the left mouse button and move the top portion of *scoop* upward, as shown in Figure 4-14.

Figure 4-13 *The scoop fitting inside the cone*

Figure 4-14 *The top portion of the scoop moved up using the **Move** brush*

8. Choose the Current Brush button from the left shelf; a flyout containing different sculpting brushes is displayed. Choose the **Standard** brush from this flyout.

9. Set the value of the **Draw Size** and the **Z Intensity** sliders to **80** and **25**, respectively. Next, press and hold the left mouse button and drag the cursor on the surface of *scoop* to create bumps in the *scoop*, as shown in Figure 4-15.

10. Again, choose the Current Brush button from the left shelf; a flyout containing different sculpting brushes is displayed. Choose the **Spiral** brush from this flyout.

11. Set the value of the **Draw Size** and the **Z Intensity** sliders to **200** and **100**, respectively. Next, press and hold the left mouse button and drag the cursor on the top of *scoop* to create its tip, refer to Figure 4-16.

*Figure 4-15 Bumps created in the scoop using the **Standard** brush*

*Figure 4-16 Shape of the scoop modified using the **Spiral** brush*

Saving the Model

In this section, you will save the file by following the steps given next.

1. Choose the **Save As** button from the **Tool** palette, the **Save ZTool** dialog box is displayed. In this dialog box, browse to the location *\Documents\ZBrushprojects\c04*.

2. Enter **c04tut1** in the **File name** edit box and then choose the **Save** button.

Tutorial 2

In this tutorial, you will create a monument using different subtools. The final output of the model is shown in Figure 4-17. **(Expected time: 45 min)**

The following steps are required to complete this tutorial:

a. Create the dome of the monument.
b. Create walls and roof of the monument.
c. Create the door of the monument.
d. Create different patterns on the walls of the monument.
e. Save the model.

Figure 4-17 The final model of the monument

Creating the Dome of the Monument

In this section, you will create the dome of the monument using a sphere and different sculpting brushes.

1. Choose the **Init ZBrush** button from the **Preferences** palette; the message box is displayed . Choose the **Yes** button from the message box; ZBrush is initialized to its default state.

2. Choose the Current Tool button from the **Tool** palette; a flyout is displayed. Choose the **Sphere3D** primitive from this flyout.

3. Press and hold the left mouse button, and drag the cursor on the canvas; a sphere is created in the canvas.

4. Choose the **Edit** button from the top shelf.

5. Choose the **PolyF** button from the right shelf; the polygon edges of the sphere are displayed.

6. Press and hold the left mouse button, and drag the cursor upward in such a way that the sphere is rotated. While dragging the cursor, keep the SHIFT key pressed; the sphere is snapped at right angle to the canvas, as shown in Figure 4-18. Choose the **PolyF** button again; the polygon edges on the sphere disappear.

7. Choose the **Make PolyMesh3D** button from the **Tool** palette to convert the primitive sphere into a polymesh.

8. In the **Geometry** subpalette of the **Tool** palette, set the value of the **SDiv** slider to **5** by clicking on the **Divide** button four times.

9. Rename the sphere to **dome** using the **Rename** button in the **SubTool** subpalette.

10. Choose the Current Brush button from the left shelf; a flyout containing different sculpting

brushes is displayed. Choose the **ClipRect** brush from this flyout; a message box is displayed, asking you to press CTRL+SHIFT keys to activate this brush. Choose the **OK** button to close this message box.

11. Press CTRL+SHIFT keys to activate the **ClipRect** brush and then press and hold the left mouse button. Next, drag the cursor over *dome*, a rectangular marquee selection appears. Select the upper half of *dome*. Now, release the left mouse button; the lower half of *dome* is deleted, refer to Figure 4-19.

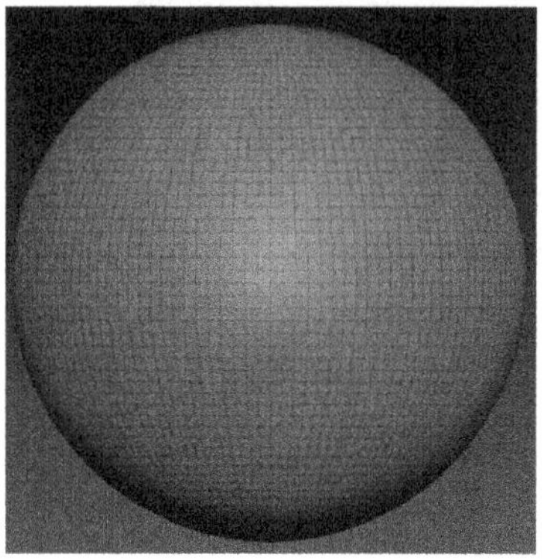

Figure 4-18 *The sphere snapped to the canvas*

Figure 4-19 *The lower half of the dome deleted*

12. Expand the **Transform** palette and then choose the **Activate Symmetry** button. Next, choose the **(R)** button; the **RadialCount** slider is activated. By default, the **>X<** button is chosen. To deactivate this button, choose it again. Next, choose the **>Y<** button; the radial symmetry in the Y-axis is activated. Set the value of the **RadialCount** slider to **60**, refer to Figure 4-20.

13. Choose the Current Brush button from the left shelf; a flyout containing different sculpting brushes is displayed. Choose the **ClayBuildup** brush from this flyout.

14. Choose the Current Stroke button from the left shelf; a flyout containing different types of strokes is displayed. Choose the **DragRect** stroke from this flyout.

15. Choose the Current Alpha button from the left shelf; a flyout containing different alpha images is displayed. Choose the **Alpha 28** alpha image from this flyout.

16. In the top shelf, set the parameters as shown in Figure 4-21.

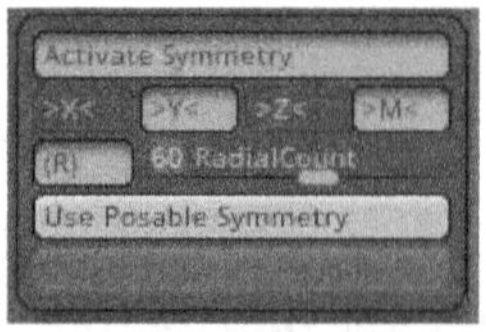

Figure 4-20 *Radial symmetry activated in Z-axis*

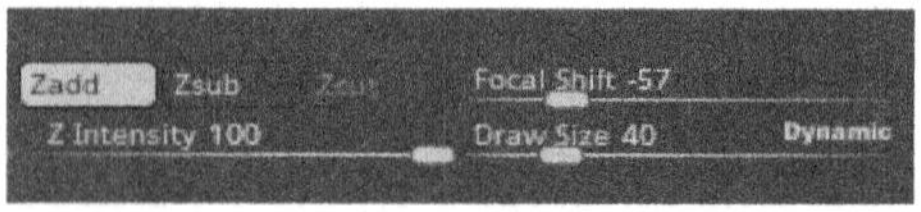

Figure 4-21 *Settings in the top shelf*

17. Press and hold the left mouse button and drag the cursor on the lower part of *dome* to create a pattern, as shown in Figure 4-22.

18. Choose the Current Brush button from the left shelf; a flyout containing different sculpting brushes is displayed. Choose the **Standard** brush from this flyout.

19. In the **Transform** palette, set the value of the **RadialCount** slider to **30**, as shown in Figure 4-23. Next, set the parameters in the top shelf, as shown in Figure 4-24.

Figure 4-22 *Pattern created on the lower part of the dome*

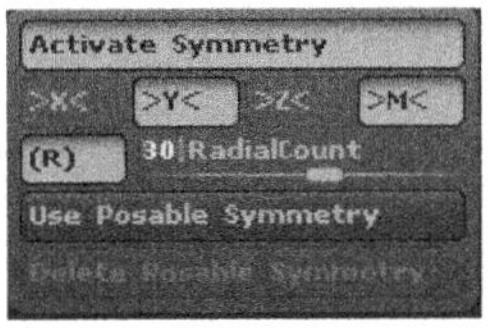

Figure 4-23 *The value of the* ***RadialCount*** *slider set to* ***30***

20. Choose the Current Stroke button from the left shelf; a flyout containing different types of strokes is displayed. Choose the **FreeHand** stroke from this flyout.

21. Press and hold the left mouse button and hover the cursor at the top of *dome*. Next, drag the cursor downward to create vertical lines on *dome*, as shown in Figure 4-25. While dragging the cursor, press and hold the SHIFT key to get straight strokes.

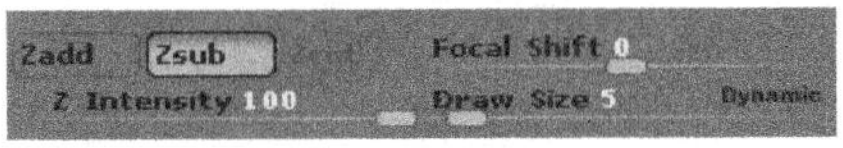

Figure 4-24 *Settings in the top shelf*

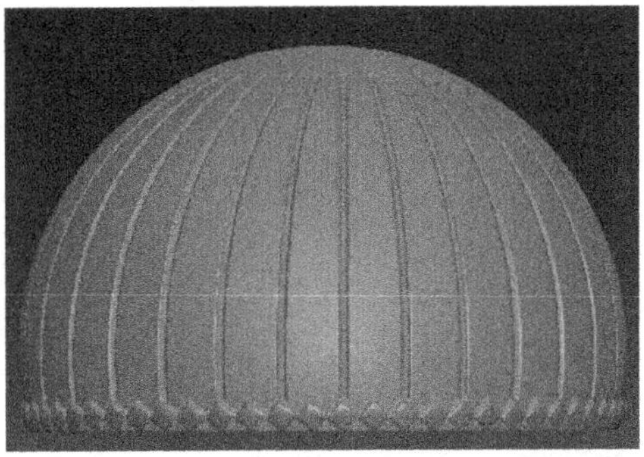

Figure 4-25 *Vertical lines created on the dome*

22. Similarly, create horizontal lines on *dome* by dragging the cursor from left to right; the tiles will be created, as shown in Figure 4-26.

23. Choose the Current Stroke button from the left shelf; a flyout containing different types of strokes is displayed. Choose the **DragRect** stroke from this flyout.

24. Choose the Current Alpha button from the left shelf; a flyout containing different alpha images is displayed. Choose the **Alpha 17** alpha image from this flyout.

25. Choose the **Zadd** button from the top shelf. Next, set the value of the **Z Intensity** slider to **30** and the **Draw Size** slider to **38**. Press and hold the left mouse button and drag the cursor at the bottom tiles of the dome to create the pattern on *dome*. Next, deactivate the symmetry.

26. In the **SubTool** subpalette, choose the **Append** button; a flyout is displayed. Choose **Sphere3D_1** from this flyout; a new sphere is created in the canvas and it overlaps the existing sphere. Make sure that you select the thumbnail for the newly created sphere from the **SubTool** list and rename it as **dome top**.

27. Make sure the **Gizmo 3D Y** button is deactivated. Choose the **Move** button from the top shelf and move **dome top** upward in such a way that it touches the top of **dome**. Next, in the **Geometry** subpalette of the **Tool** palette, set the value of the **SDiv** slider to **5** by clicking on the **Divide** button four times.

28. Expand the **Deformation** subpalette in the **Tool** palette. In this subpalette, move the **Size** slider toward the left so that the size of *dome top* decreases and it fits into the top of *dome*. Next, move *dome top* in such a way that its lower half portion is inserted into *dome*, refer to Figure 4-27.

Figure 4-26 *Tiles created on the dome*

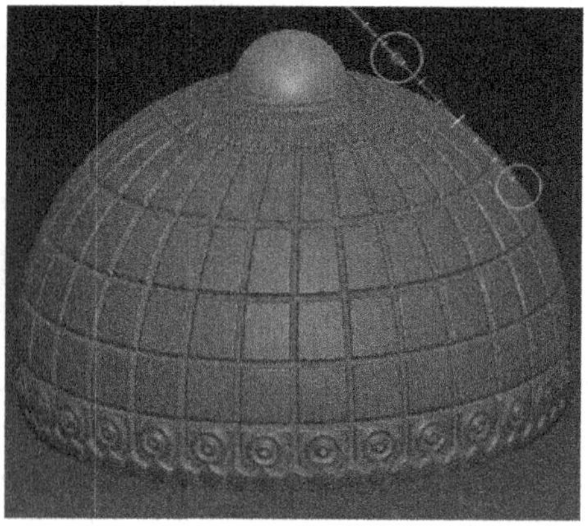

Figure 4-27 *The dome top scaled and moved to the top*

29. Make sure the **Draw** button is chosen. Choose the Current Stroke button from the left shelf; a flyout containing different types of strokes is displayed. Choose the **Dot** stroke from this flyout.

30. Choose the Current Alpha button from the left shelf; a flyout containing different alpha images is displayed. Choose **Alpha off** from this flyout.

31. Choose the **Draw** button from the top shelf. Next, choose the **Activate Symmetry** button from the **Transform** palette. Next, choose the **(R)** button; the **RadialCount** slider is activated. By default, the **>X<** button is chosen. To deactivate this button, choose it again. Next, choose the **>Y<** button; the radial symmetry in the Y-axis is activated. Set the value **40** in the **RadialCount** slider .

32. Set the value of the **Z Intensity** slider to **40** and the **Draw Size** slider to **32**. Next, press and hold the left mouse button and drag the cursor on the surface of *dome top* to create a radial pattern, as shown in Figure 4-28.

    Next, you will create the finial for the monument using the **CurveLathe** brush.

33. Before using the **CurveLathe** brush on *dome top*, you have to make sure all lower division levels are deleted. To do so, choose the **Del Lower** button from the **Geometry** subpalette of the **Tool** palette, as shown in Figure 4-29.

Figure 4-28 *Pattern created on the dome top using radial symmetry*

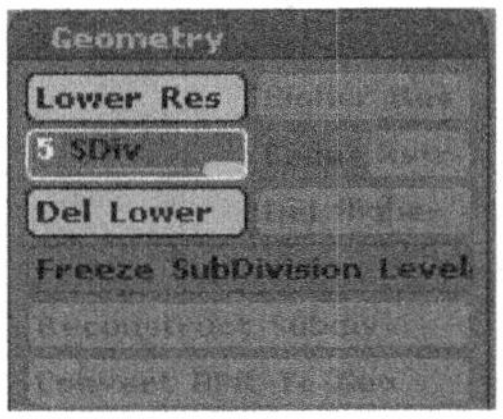

Figure 4-29 *The **Del Lower** button chosen in the **Geometry** subpalette*

34. Choose the **(R)** and **>Y<** buttons in the **Transform** palette; the radial symmetry in the Y-axis is deactivated.

35. Choose the Current Brush button from the left shelf; a flyout containing different sculpting brushes is displayed. Choose the **CurveLathe** brush from this flyout. Next, press and hold the left mouse button and draw a profile curve at the top of *dome top*, refer to Figure 4-30. Release the left mouse button; a finial is created on the surface of *dome top*, as shown in Figure 4-30.

Creating Walls and the Roof of the Monument

In this section, you will create the walls, roof, and pillars of the monument using the cylinder primitive and different sculpting brushes.

1. Choose the **Append** button in the **SubTool** subpalette; a flyout is displayed. Choose the **Cylinder3D** primitive from this flyout; a cylinder is created in the canvas and it overlaps the *dome*, as shown in Figure 4-31. The thumbnail for the cylinder is displayed in the **SubTool** list. Make sure that you select the thumbnail for the cylinder in the list. Next, rename it to **walls**.

Figure 4-30 *Finial created on the dome top using the **CurveLathe** brush*

Figure 4-31 *Cylinder overlapping the dome*

2. Expand the **Geometry** subpalette. In this subpalette, click on the **Reconstruct Subdiv** button twice, as shown in Figure 4-32; a low resolution model of *walls* having less number of segments is displayed.

3. Make sure the **Gizmo 3D Y** button is deactivated. Choose the **Move** button from the top shelf; an action line is displayed. Move *walls* using the action line in such a way that *walls* lie below *dome*, as shown in Figure 4-33.

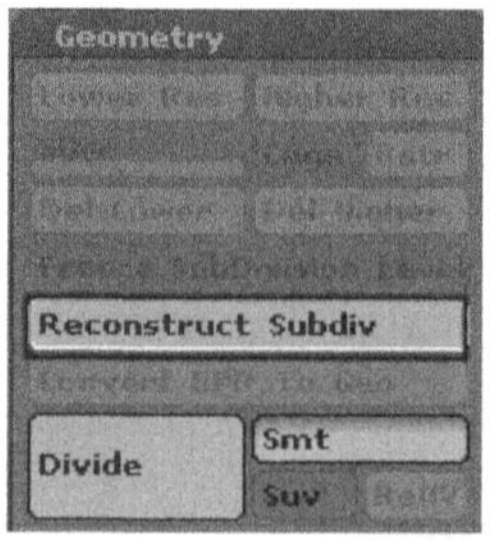

Figure 4-32 ***Reconstruct Subdiv*** *button chosen from the **Geometry** subpalette*

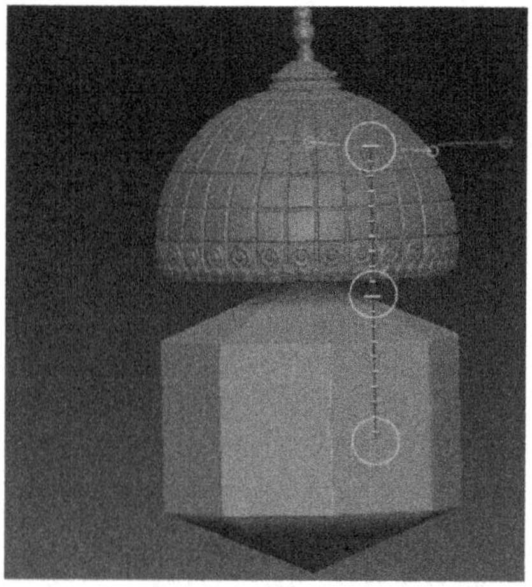

Figure 4-33 *The walls moved below the dome*

4. Expand the **Deformation** subpalette in the **Tool** palette.

You need to increase the height of *walls* along the Z-axis using the **Size** slider.

5. Deactivate the x and y options corresponding to the **Size** slider by choosing the **x** and **y** buttons located on the right side of the slider.

6. Set the value of the **Size** slider to **55** by dragging the slider toward right; the height of the cylinder increases along the Z-axis. Alternatively, you can enter the value **55** in the edit box displayed on clicking on the slider.

7. Choose the **Draw** button from the top shelf. Choose the Current Brush button from the left shelf; a flyout containing different sculpting brushes is displayed. Choose the **ClipRect** brush from this flyout; a dialog box is displayed, asking you to press CTRL+SHIFT keys to activate this brush. Choose the **OK** button to close this dialog box.

8. Press CTRL+SHIFT keys to activate the **ClipRect** brush and then press and hold the left mouse button. Next, drag the cursor over the cylinder, a rectangular marquee selection appears. Make a marquee selection on *walls*, as shown in Figure 4-34. Now, release the left mouse button; the lower and upper parts are deleted.

9. Make sure the **Gizmo 3D Y** button is deactivated. Choose the **Move** button from the top shelf and move *walls* using the action line, as shown in Figure 4-35.

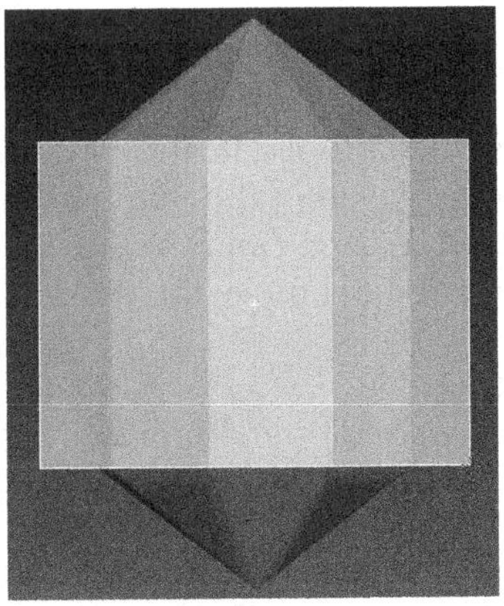

Figure 4-34 *Marquee selection on the walls*

Figure 4-35 *The walls moved under the dome*

10. Choose the **Append** button in the **SubTool** subpalette; a flyout is displayed. Choose **Cylinder3D** from this flyout; a cylinder is created in the canvas. Select the thumbnail for the newly created cylinder from the **SubTool** list. Rename it to **roof**.

11. Expand the **Deformations** subpalette in the **Tool** palette. Decrease the size of *roof* along the Z-axis by choosing the **z** button and moving the **Size** slider toward left. Similarly, increase the size of *roof* along the X and Y axes. Make sure the **Move** button is chosen from the top shelf. Now, using the action line, move *roof* below *dome,* refer to Figure 4-36.

12. Choose *walls* from the **SubTool** list. Expand the **Deformation** subpalette, and increase the size of *walls* along the X and Y axes, as shown in Figure 4-37. Make sure the **Move** button is chosen and the **Gizmo 3D Y** button is deactivated from the top shelf. Now, using the action line, move *walls* below *roof.*

Figure 4-36 *The roof scaled and moved below the dome*

Figure 4-37 *The size of the walls increased along X and Y axes*

13. In the **SubTool** subpalette, select *roof* from the list of subtools. In this subpalette, choose the **Duplicate** button; a copy of *roof* is created and displayed in the **SubTool** list. Select the duplicate *roof* from the list. Make sure the **Move** button is chosen and the **Gizmo 3D Y** button is deactivated from the top shelf. Now, using the action line, move the duplicate *roof* below *walls.*

14. Expand the **Deformation** subpalette and increase the size of the duplicate *roof* along the X and Y axes, refer to Figure 4-38.

15. In the **SubTool** subpalette, select *roof* from the list of subtools. Next, choose the **Activate Symmetry** button from the **Transform** palette. Choose the **(R)** button; the **RadialCount** slider is activated. By default, the **>X<** button is chosen. To deactivate this button, choose it again. Next, choose the **>Z<** button; the radial symmetry in the Z-axis is activated. In the **RadialCount** edit box, enter the value **46**, refer to Figure 4-39.

16. Choose the **Draw** button from the top shelf, if not already chosen and then choose the **InsertHRing brush** from the **Brush** tab of the **LightBox** browser

17. Choose the Current Stroke button from the left shelf; a flyout containing different types of strokes is displayed. Choose the **Dots** stroke from this flyout, if not already chosen.

Figure 4-38 *Duplicate roof scaled along the X and Y axes*

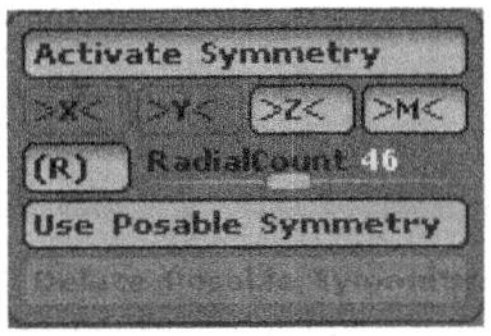

Figure 4-39 *The value of **RadialCount** slider set to **46***

18. In the top view, press and hold the left mouse button and drag the cursor at the corners of the *roof* to create the railings, refer to Figure 4-40.

    Next, you will create pillars for the monument using the **CurveLathe** brush.

19. Choose the **(R)** and **>Z<** buttons in the **Transform** palette; the radial symmetry in the Z-axis is deactivated.

20. Choose the Current Brush button from the left shelf; a flyout containing different sculpting brushes is displayed. Choose the **CurveLathe** brush from this flyout. Next, press and hold the left mouse button and draw a profile curve below *roof*. Release the left mouse button; a pillar is created, refer to Figure 4-41.

Figure 4-40 *Railing created using the* ***InsertHRing*** *brush*

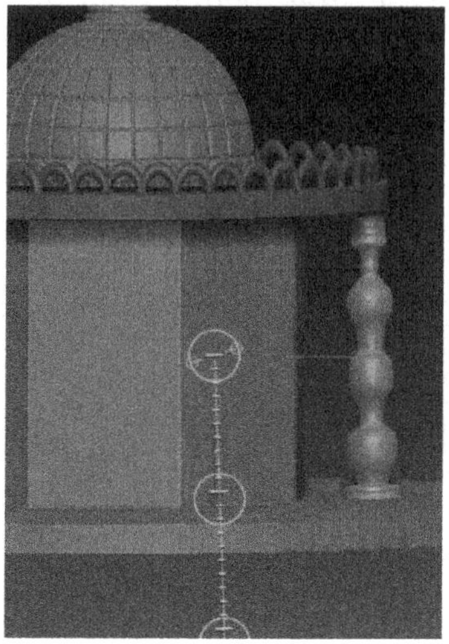

Figure 4-41 *Pillar created using the* ***CurveLathe*** *brush*

21. Make sure the **Gizmo 3D Y** button is deactivated. Choose the **Move** button from the top shelf; an action line is displayed. Press and hold the left mouse button and drag the cursor vertically to increase the size of the action line. Next, press the CTRL key and move the pillar on the left side using the action line; a duplicate copy of the pillar is displayed. Next, move the duplicate pillar to the other side of *walls* using the action line, refer to Figure 4-42.

Creating the Door of the Monument

In this section, you will create the door of the monument using a plane and the **Projection Master** dialog box.

1. Choose the **Append** button in the **SubTool** subpalette; a flyout is displayed. Choose **Plane3D** from this flyout; a plane is created in the canvas. The thumbnail for the plane is displayed in the **SubTool** list. Make sure that you select the thumbnail for the plane in the list. Next, rename it to **door**.

2. Expand the **Deformations** subpalette in the **Tool** palette. You need to rotate *door* along the Y-axis using the **Rotate** slider.

3. Deactivate the x and z options corresponding to the **Rotate** slider by choosing the **x** and **z** buttons located on the right side of the slider. Next, activate the y option by choosing the **y** button.

4. Set the value of the **Rotate** slider by entering the value **-90** in the edit box displayed on clicking on the slider; *door* is rotated along the Y-axis.

5. Press and hold the left mouse button and drag the cursor in the canvas toward the left; the side view is displayed. Make sure the **Gizmo 3D Y** button is deactivated. Choose the **Move** button from the top shelf. Next, move the *door* to the front of *walls* using the action line. Next, switch back to the front view by dragging the cursor toward right.

6. In the **Deformation** subpalette of the **Tool** palette, deactivate the x option corresponding to the **Size** slider by choosing the **x** button located on the right side of the slider. Make sure the **y** and **z** buttons are chosen. Decrease the height and width of *door* by dragging the **Size** slider toward left. Next, choose the **Move** button from the top shelf and position *door*, as shown in Figure 4-43.

7. Expand the **Geometry** subpalette in the **Tool** palette. In this subpalette, click on the **Divide** button four times; the value of the **SDiv** slider becomes **5**.

8. Choose the **ZPlugin** palatte; a flyout is displayed. Expand **Projection Master** from the flyout and then choose the **Projection Master** button from the top shelf; the **Projection Master** dialog box is displayed.

9. Select the **Deformation** check box from the **Projection Master** dialog box and make sure that the other check boxes are not selected, refer to Figure 4-44. Next, choose the **DROP NOW** button from this dialog box; the 3D plane is converted into its 2.5D illustration.

Figure 4-42 *Copy of the pillar created and moved toward left*

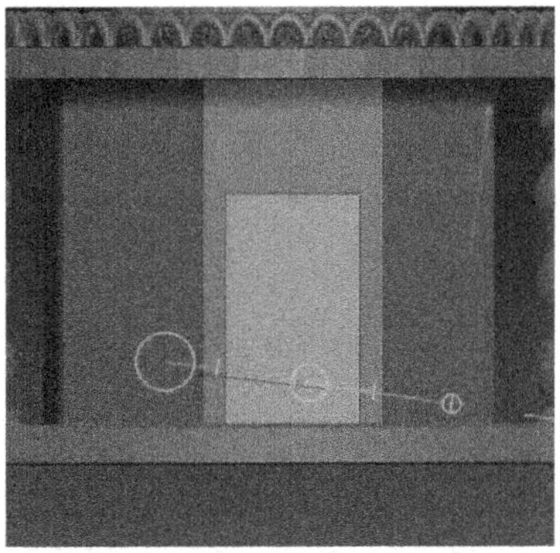

Figure 4-43 *The door moved using the action line*

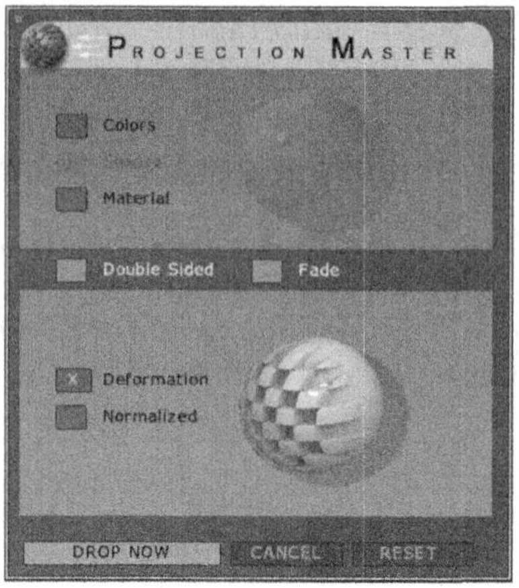

Figure 4-44 *The* ***Deformation*** *check box selected*

10. Choose the Current Stroke button from the left shelf; a flyout containing different types of strokes is displayed. Choose the **LineII** stroke from this flyout.

11. Choose the Current Alpha button from the left shelf; a flyout containing different alpha images is displayed. Choose the **Alpha 28** alpha image from this flyout.

12. Choose the **Zsub** button from the top shelf and set the value of the **Draw Size** slider to **2**. Next, press and hold the left mouse button and drag it to create the horizontal and vertical partitions of the *door*, as shown in Figure 4-45.

13. Choose the Current Stroke button from the left shelf; a flyout containing different types of strokes is displayed. Choose the **DragRect** stroke from this flyout. Next, press and hold the left mouse button and drag the cursor to create a pattern on top left panel of the door.

14. Choose the **Scale** button from the top shelf; a gyro is displayed. Hover the cursor on the pink ring of the gyro. Next, drag the cursor upward and release the left mouse button; the pattern is scaled up, refer to Figure 4-46.

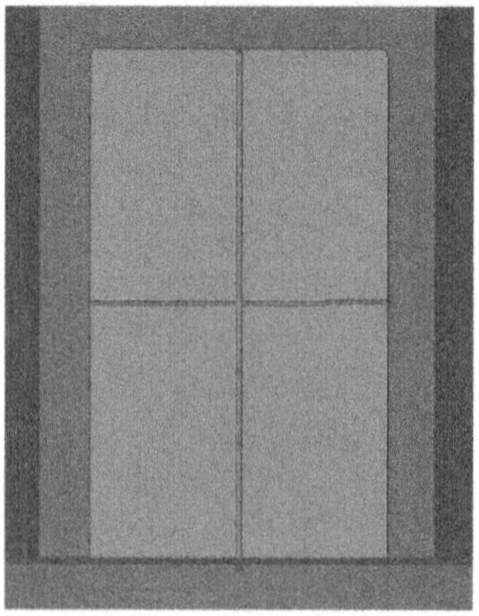

Figure 4-45 *Horizontal and vertical partitions created on the door*

Figure 4-46 *The pattern scaled up*

15. Make sure the **Gizmo 3D Y** button is deactivated. Choose the **Move** button from the top shelf; a gyro is displayed. Press and hold the left mouse button, and then hover the cursor on the pink ring of the gyro. If required, drag the cursor upward to move the pattern to the centre of the panel.

16. Press SHIFT+S to create a copy of the pattern and then press and hold the left mouse button. Now, hover the cursor on the pink ring of the gyro. Next, drag the cursor downward to move the copy of the pattern to the centre of the bottom left panel. Next, release the left mouse button and SHIFT+S.

17. Make sure the **Move** button is chosen in the top shelf. Press SHIFT+S and then press and hold the left mouse button. Now, hover the cursor on the blue ring of the gyro. Next, drag the cursor toward the right to move the copy of the pattern to the centre of the bottom right panel. Next, release the left mouse button and SHIFT+S.

18. Press SHIFT+S and then press and hold the left mouse button. Now, hover the cursor on the pink ring of the gyro. Next, drag the cursor upward to move the copy of the pattern to the centre of the top right panel. Next, release the left mouse button and SHIFT+S; the pattern is created on all the panels, refer to Figure 4-47.

19. Choose the **Draw** button and then choose the **Zadd** button from the top shelf. Next, press and hold the left mouse button and drag the cursor on any of the existing patterns; another pattern is created on the existing pattern. Scale and move the pattern such that it fits into the existing pattern. Copy the newly created pattern on all the panels, refer to Figure 4-48.

Figure 4-47 *Pattern created on all the panels*

Figure 4-48 *Newly created pattern copied on all the panels*

20. After copying the patterns, choose the **Draw** button from the top shelf. Next, choose the Current Alpha button from the left shelf; a flyout containing different alpha images is displayed. Choose the **Alpha 28** alpha image from this flyout.

21. Set the value of the **Draw Size** and the **Z Intensity** sliders to **25** and **40**, respectively. Next, press and hold the left mouse button and drag the cursor at the centre of any of the panels; a pattern is created. Copy the newly created pattern to all the panels, as shown in Figure 4-49.

22. Choose the Current Alpha button from the left shelf; a flyout containing different alpha images is displayed. Choose the **Alpha 54** alpha image from this flyout.

23. Set the value of the **Draw Size** and the **Z Intensity** sliders to **17** and **100**, respectively. Next, press and hold the left mouse button and drag the cursor at the centre of *door*; a knob is created, refer to Figure 4-50.

Figure 4-49 *Pattern created on all the panels*

Figure 4-50 *Knob of the door created*

24. Choose the **Projection Master** button from the top shelf; the **Projection Master** dialog box is displayed. Next, choose the **PICKUP NOW** button from this dialog box; the 2.5D illustration is converted back into a 3D model.

Creating Different Patterns on Walls of the Monument

In this section, you will create different patterns on the walls of the monument using the **Projection Master**.

1. Choose *walls* from the **SubTool** list. Next, expand the **Geometry** subpalette. In this subpalette, choose the **Del Higher** button; all subdivision levels above the current **SDiv** value are deleted.

2. Choose the **Smt** button in the **Geometry** subpalette; it will be deactivated. If the **Smt** button is deactivated, the hard edges of the model will be visible even if you subdivide the geometry. Next, click on the **Divide** button seven times; the value of the **SDiv** slider is set to **8**.

3. Choose the **Projection Master** button from the top shelf; the **Projection Master** dialog box is displayed. Select the **Deformation** check box from this dialog box, and make sure that the other check boxes are cleared. Next, choose the **DROP NOW** button from this dialog box.

4. Choose the Current Stroke button from the left shelf; a flyout containing different types of strokes is displayed. Choose the **Grid** stroke from this flyout.

5. Choose the Current Alpha button from the left shelf; a flyout containing different alpha images is displayed. Choose the **Alpha 28** alpha image from this flyout.

6. Choose the **Zsub** button from the top shelf. Next, press and hold the left mouse button, and drag the cursor on the top of *door*; a grid pattern is created on *walls*, as shown in Figure 4-51.

7. Similarly, create different patterns on *walls* using different strokes and alpha images, refer to Figure 4-54.

8. Choose the **Projection Master** button from the top shelf again; the **Projection Master** dialog box is displayed. Next, choose the **PICKUP NOW** button from this dialog box; the 2.5D illustration is converted back into a 3D model, refer to Figure 4-52.

Figure 4-51 *Grid pattern created on the walls*

Figure 4-52 *The final model of the monument*

Saving the Model

In this section, you will save the file using the steps given next.

1. Choose the **Save As** button from the **Tool** palette, the **Save ZTool** dialog box is displayed. In this dialog box, browse to the location *\Documents\ZBrushprojects\c04*.

2. Enter **c04tut2** in the **File name** edit box and then choose the **Save** button.

EXERCISES

The output of the models used in the following exercises can be accessed by downloading the *c04_ZBrush_2018_exr.zip* file from *www.cadsofttech.com*. The path of the file is as follows: *Textbooks > Animation and Visual Effects > ZBrush > Pixologic ZBrush 2018 for Novices*.

Exercise 1

Create the model of a house using different subtools, as shown in Figure 4-53.

(Expected time: 45 min)

Figure 4-53 *Model of a house*

Exercise 2

Create the scene using different subtools, refer to Figure 4-54. **(Expected time: 20 min)**

Figure 4-54 *Scene to be created for Exercise 2*

Chapter 5

ZSpheres

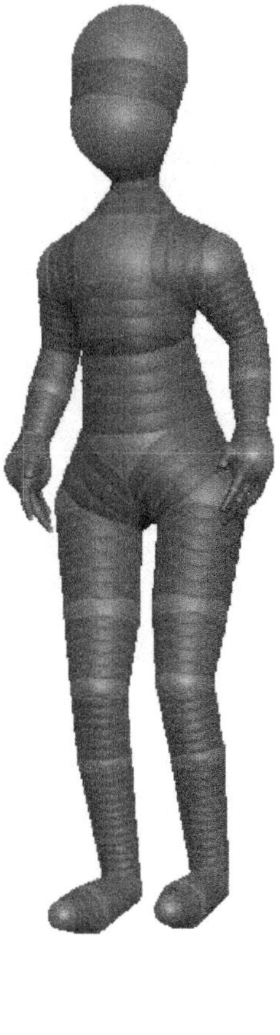

Learning Objectives

After completing this chapter, you will be able to:

- *Work with ZSpheres*
- *Understand ZSketching*
- *Create basic models using ZSpheres*

INTRODUCTION

ZSpheres are the modeling tools that can be connected to a network to create a basic structure of a model. You can access ZSpheres from the **Tool** palette. These tools are commonly used in organic modeling. In this chapter, you will learn about the process of creating an armature using ZSpheres.

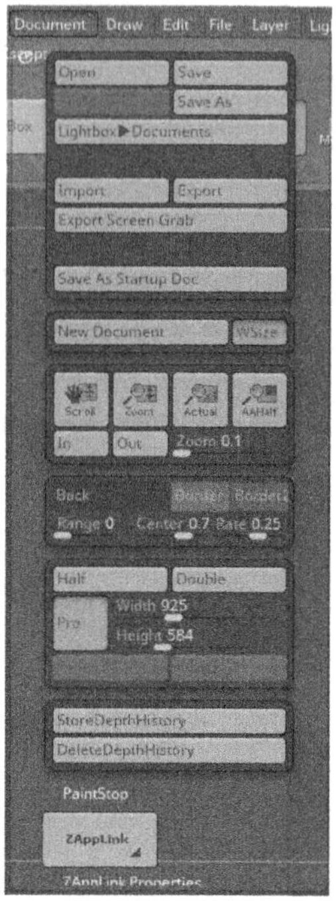

Figure 5-1 *The value of the* ***Range*** *slider set to 0*

Note

In this chapter, the background color of the canvas has been changed to white to help you to view the images better. To change the background color of the canvas, set the color in both the ***Secondary Color*** *and* ***Main Color*** *swatches in the left shelf to white. Next, choose the* ***Document*** *palette. In this palette, set the value of the* ***Range*** *slider to* ***0****, refer to Figure 5-1. The* ***Range*** *slider is used to add the gradient effect. If its value is set to* ***0****, no gradient will be displayed. Choose the* ***Back*** *button in this palette; the color of the canvas will change to white.*

TUTORIAL

All the files used in the tutorial can be downloaded from the CADSoft website (*www.cadsofttech.com*).These files are compressed in zip file format and are required to be extracted before using them in the tutorials. The path of the files is as follows: *Textbooks > Animation and Visual Effects > Pixologic ZBrush 2018 > Pixologic ZBrush 2018 for Novices*

Tutorial 1

In this tutorial, you will create the basic shape of a human body using ZSpheres. The final output of the model is shown in Figure 5-2. **(Expected time: 30 min)**

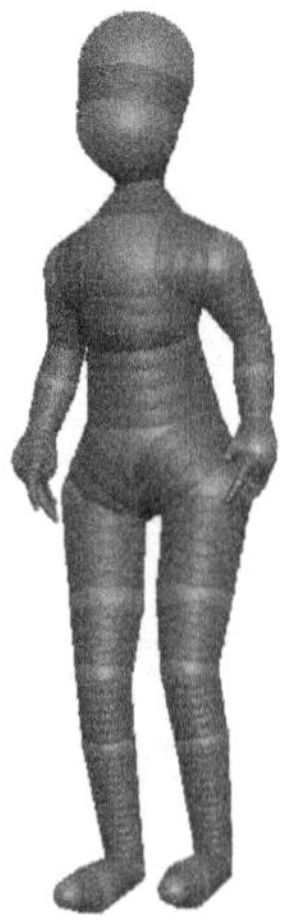

Figure 5-2 *Basic shape of the human body*

The following steps are required to complete this tutorial:

a. Create the torso of the human body.
b. Create the limbs.
c. Create the neck and the head.
d. Save the model.

Creating the Torso of the Human Body

In this section, you will create the torso of the human body using a framework of ZSpheres.

1. Choose the Current Tool button from the **Tool** palette; a flyout is displayed. In this flyout, choose the **ZSphere** primitive and create it on the canvas, as shown in Figure 5-3. Next, choose the **Edit** button from the top shelf.

2. Activate the symmetry in the X-axis by pressing the X key. Hover the cursor at the centre of the bottom portion of the ZSphere; a green circle is displayed, as shown in Figure 5-4.

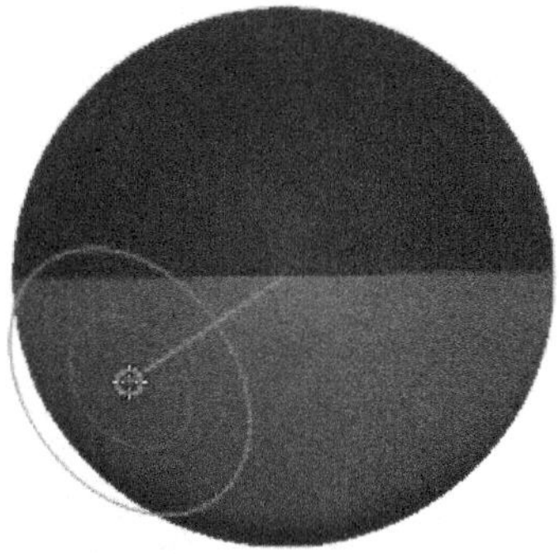

Figure 5-3 *ZSphere created in the canvas*

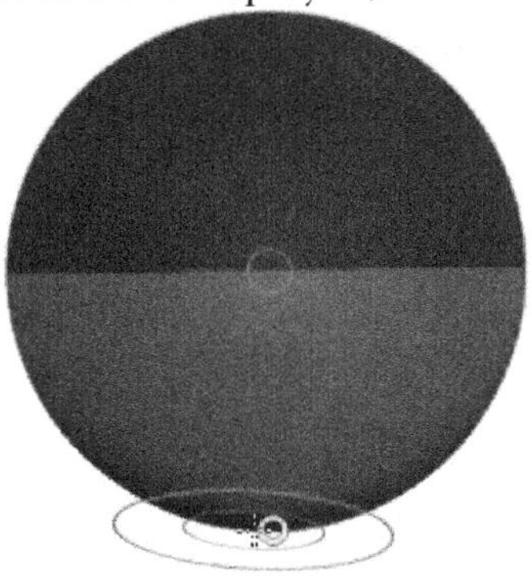

Figure 5-4 *Green circle displayed*

3. Create a second ZSphere at the position where the green circle is displayed, as shown in Figure 5-5.

4. Similarly, create a third ZSphere at the centre of the top portion of the ZSphere, refer to Figure 5-6. If required, choose the **Move** button from the top shelf and position the ZSpheres such that they lie in a straight line.

5. Choose the **Scale** button from the top shelf. Next, select the middle ZSphere by clicking on it. Press and hold the left mouse button and drag the cursor inward to scale down the ZSphere, as shown in Figure 5-7.

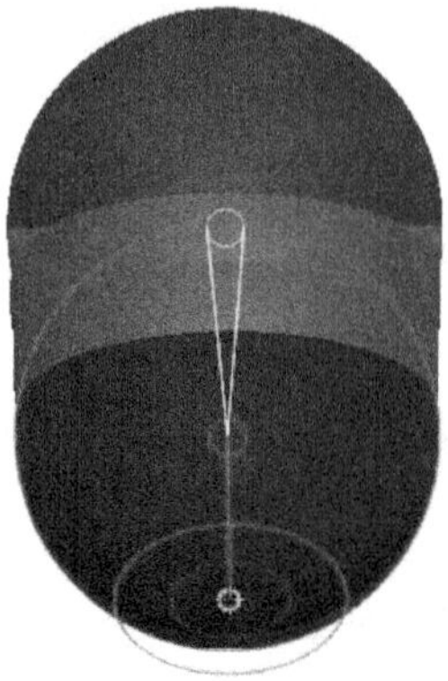

Figure 5-5 Second ZSphere created below the existing ZSphere

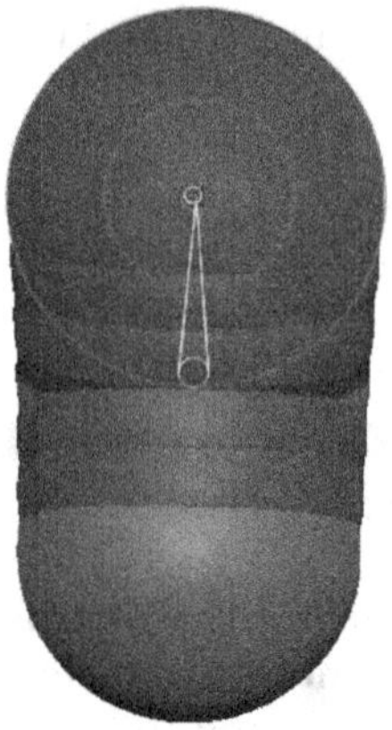

Figure 5-6 Third ZSphere created above the existing ZSphere

Creating the Limbs

In this section, you will create the hands and legs of the human body by creating new ZSpheres on the existing structure.

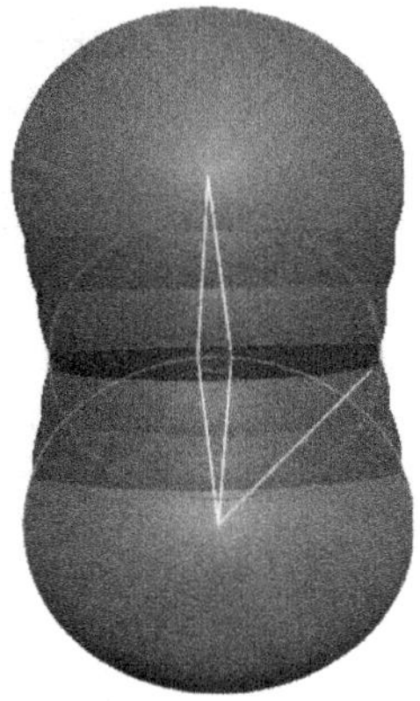

*Figure 5-7 Middle most ZSphere scaled using the **Scale** button*

1. Choose the **Draw** button from the top shelf. Next, hover the cursor on the right side of the bottom ZSphere and create a ZSphere for the hips. As the symmetry is activated in the X-axis, a ZSphere is created also on the left side, refer to Figure 5-8.

2. Press and hold the left mouse button and drag the cursor in the canvas area toward the right till the side view of the structure is displayed. Next, move the newly created ZSpheres slightly backwards and position them, as shown in Figure 5-9. Again, switch back to the front view by dragging the cursor in the canvas area toward the left.

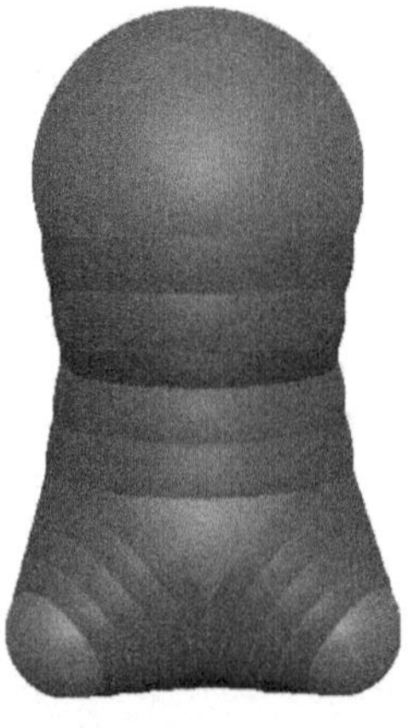

Figure 5-8 ZSphere created for the hips

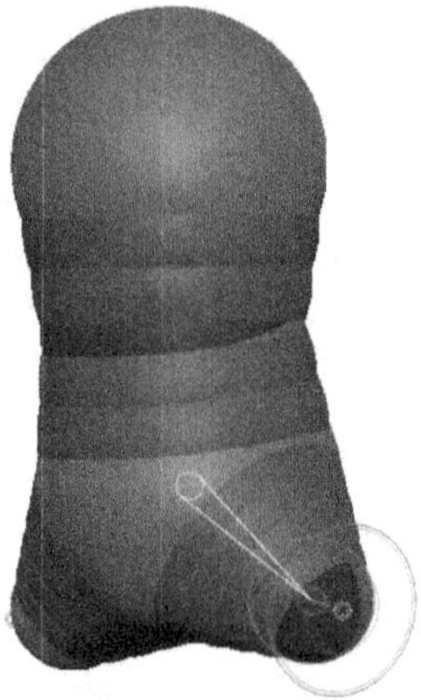

Figure 5-9 Positioning ZSphere in the side view

3. Create the ZSpheres for the shoulders of the body, refer to Figure 5-10. Position the shoulders in the side view, as discussed in Step 2.

4. Hover the cursor on the ZSphere created for the hips of the body. Next, create a new ZSphere on it and move it downwards, as shown in Figure 5-11.

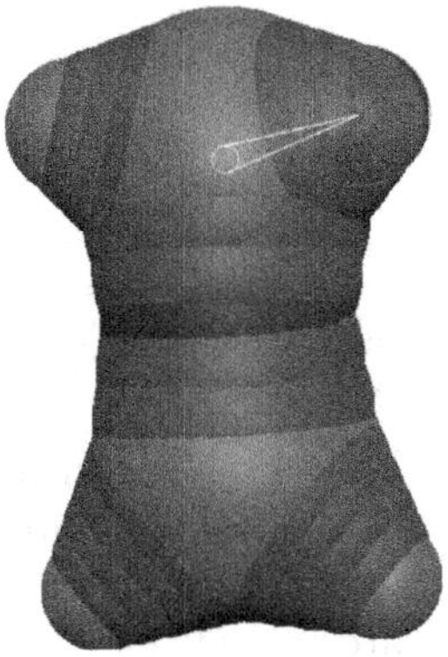

Figure 5-10 *ZSphere created for the shoders*

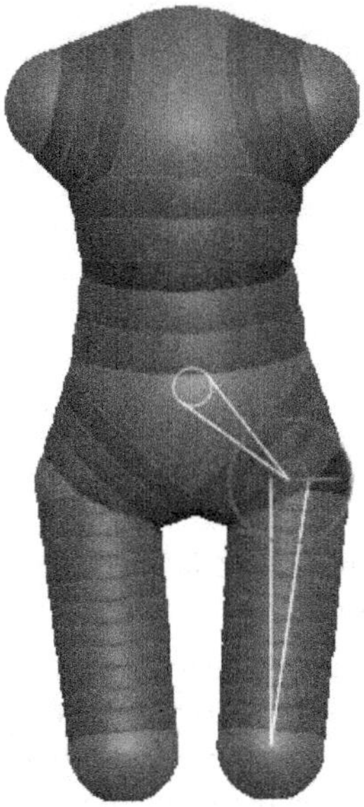

Figure 5-11 *ZSphere created and moved downward*

5. Create new ZSpheres for the arms and legs of the body, and move them downward, refer to Figure 5-12.

6. Make sure the **Draw** button is chosen in the top shelf. Next, insert two editable ZSpheres on the legs by clicking on two linked ZSpheres, refer to Figure 5-13. Choose the **Scale** button from the top shelf and scale up the inserted ZSpheres, as shown in Figure 5-13.

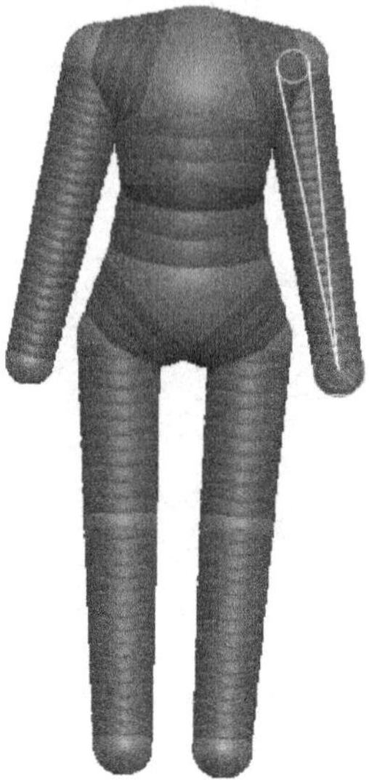

Figure 5-12 ZSpheres created for the legs and the arms

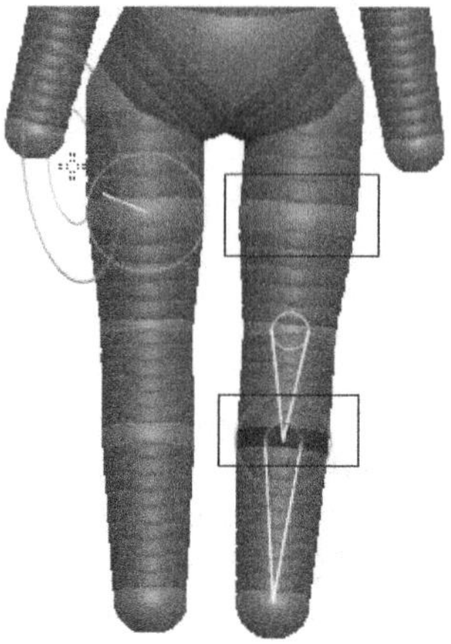

Figure 5-13 Two new ZSpheres inserted and scaled

7. Press and hold the left mouse button and drag the cursor in the canvas area toward the right till the side view of the structure is displayed. Next, choose the **Move** button from the top shelf and move the ZSpheres on the legs to position them, as shown in Figure 5-14. Switch back to the front view by dragging the cursor toward the left.

8. Choose the **Draw** button from the top shelf and create two new ZSpheres for the feet of the body. Next, choose the **Move** button and move the newly created ZSpheres toward the front. Choose the **Scale** button and scale up the middle ZSphere, as shown in Figure 5-15.

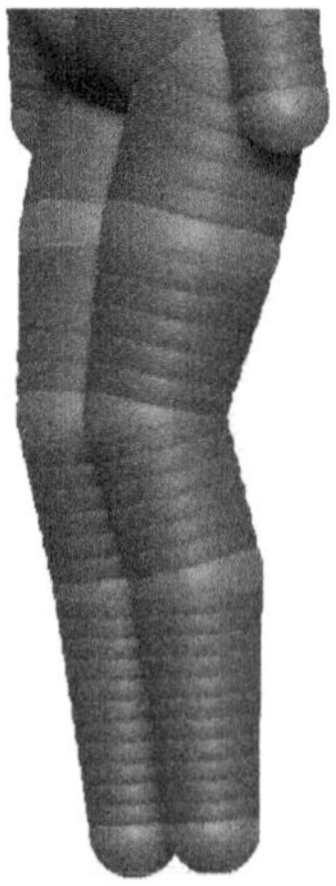

Figure 5-14 The ZSpheres adjusted in the side view

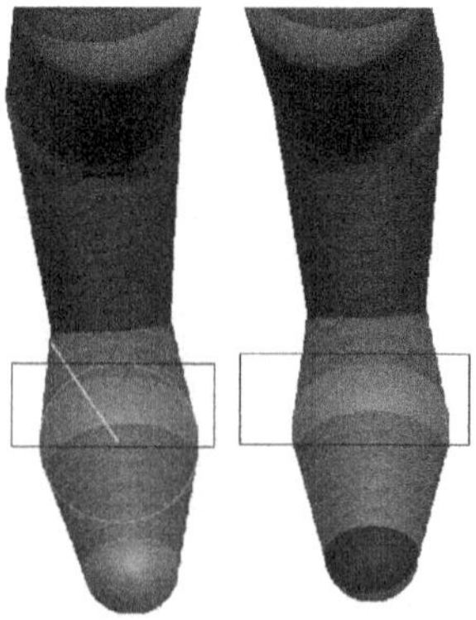

Figure 5-15 Middle ZSphere scaled up

9. Insert two editable ZSpheres on the arms by clicking on the two linked ZSpheres, refer to Figure 5-16.

10. Press and hold the left mouse button and drag the cursor in the canvas area toward the right till the side view of the structure is displayed. Next, move the inserted ZSpheres slightly backward in the side view to create a bend in the elbow, refer to Figure 5-16.

11. Create a new ZSphere for the palm of the hand, and move it downward, as shown in Figure 5-17.

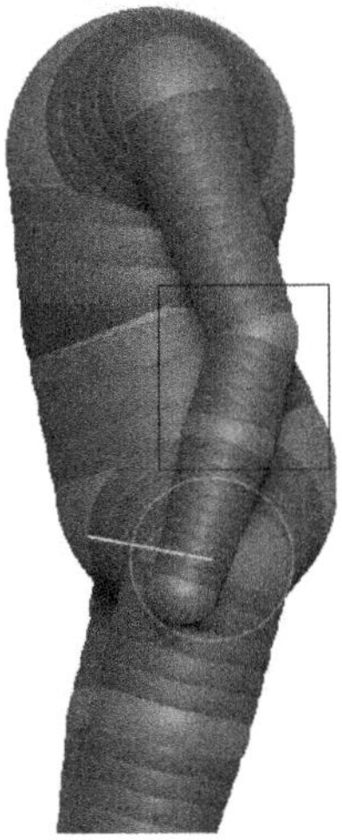

Figure 5-16 *Two new ZSpheres inserted and positioned in the side view*

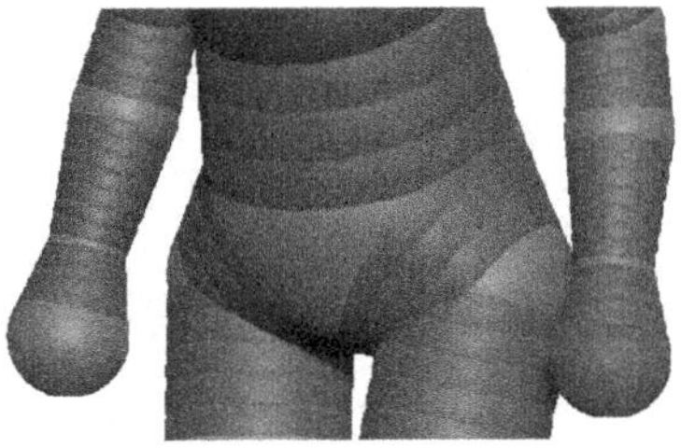

Figure 5-17 *New ZSphere created for the palm and moved downwards*

12. Choose the **Move** button from the top shelf and select the ZSphere created for the palm. You will notice that its color changes. Hover the cursor at the area that divides the red and maroon tones of the ZSphere, as shown in Figure 5-18. Next, choose the **Draw** button and create a ZSphere for the knuckle of thumb and move it outward, as shown in Figure 5-19.

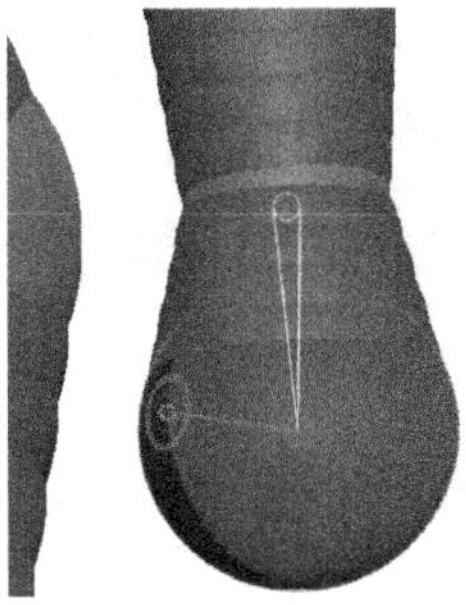

Figure 5-18 *Cursor hovered at the area dividing the red and maroon tones of the ZSphere*

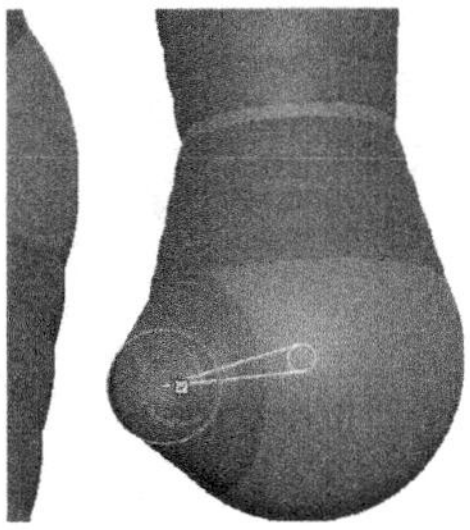

Figure 5-19 *ZSphere created and moved outward*

13. Create four more ZSpheres for the knuckles of the rest of the fingers and move them outward, as shown in Figure 5-20.

14. Create new ZSpheres for the thumb and the index finger and then move them outward, as shown in Figure 5-21.

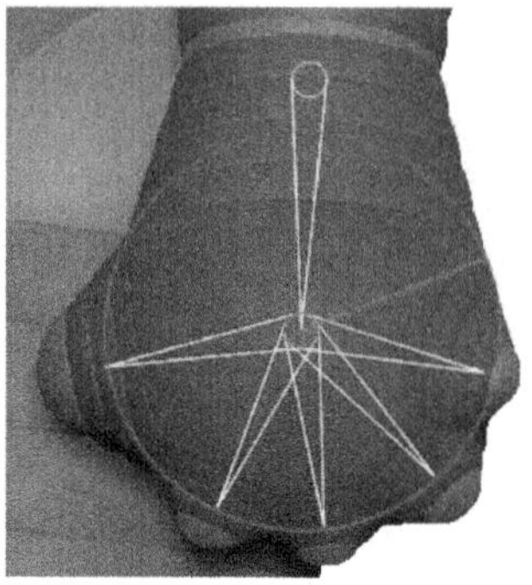

Figure 5-20 *Knuckles created for all the fingers of the hand*

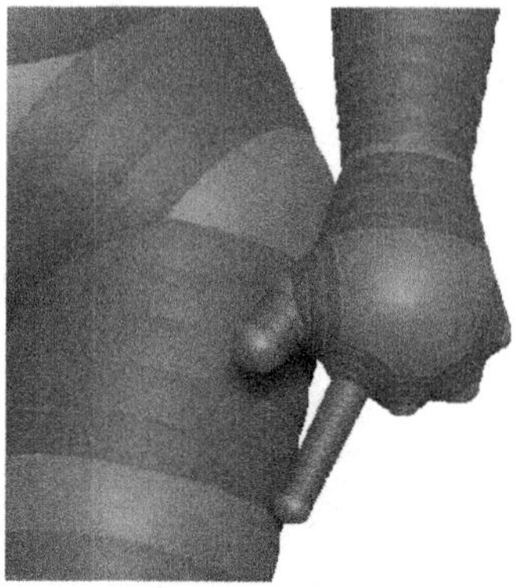

Figure 5-21 *Thumb and index finger created by creating ZSpheres and moving them outward*

15. Create the middle finger by creating a ZSphere at the knuckle and then moving it outward. Next, insert a ZSphere at the middle of the index finger and the middle finger by clicking on the linked spheres, as shown in Figure 5-22.

16. Similarly, create the ring finger and little finger and adjust them by scaling and moving the ZSpheres in them, refer to Figure 5-23.

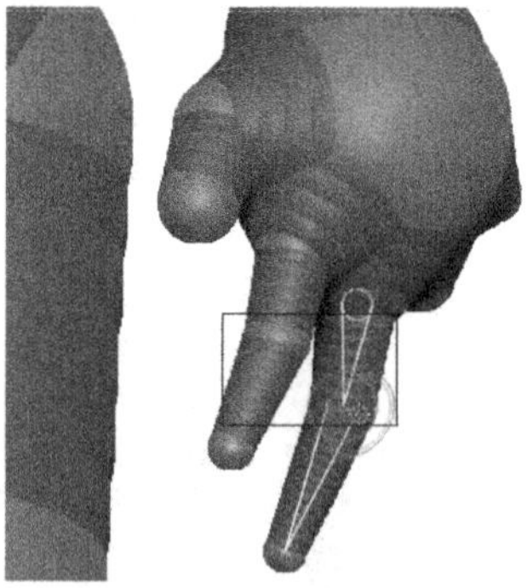

Figure 5-22 *Two new ZSpheres inserted in the index finger and the middle finger*

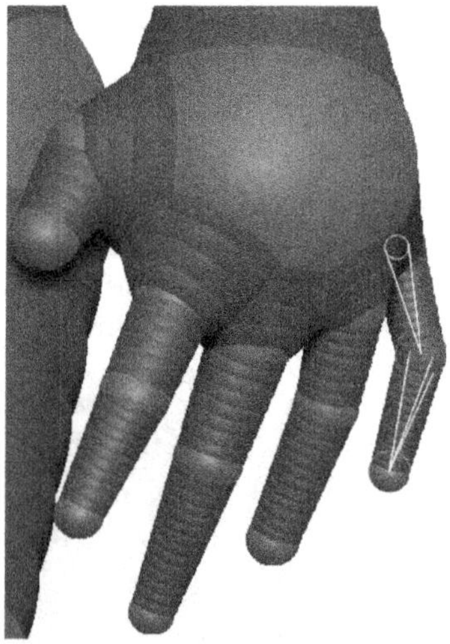

Figure 5-23 *Ring finger and the little finger created using ZSpheres*

Creating the Neck and the Head

In this section, you will create the neck and head of the human body by creating new ZSpheres on the existing structure.

1. Choose the **Move** button from the top shelf. Hover the cursor on the top-most ZSphere of the framework and place it at the middle of the area that divides the red and maroon tones of the ZSphere. Next, choose the **Draw** button, and create a ZSphere for the lower part of the neck, refer to Figure 5-24.

2. Create another ZSphere for the upper part of the neck, refer to Figure 5-25.

3. Create a ZSphere on the top of the neck for the lower part of the head, refer to Figure 5-26.

4. Similarly, create another ZSphere for the upper part of the head, refer to Figure 5-27.

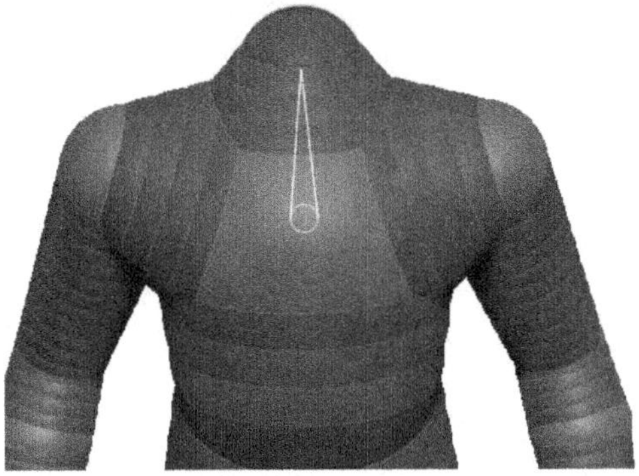

Figure 5-24 *ZSphere created for the lower part of the neck*

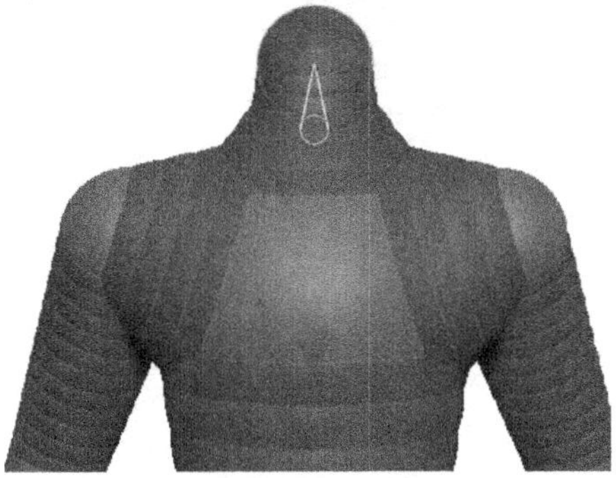

Figure 5-25 *Second ZSphere created for the upper part of the neck*

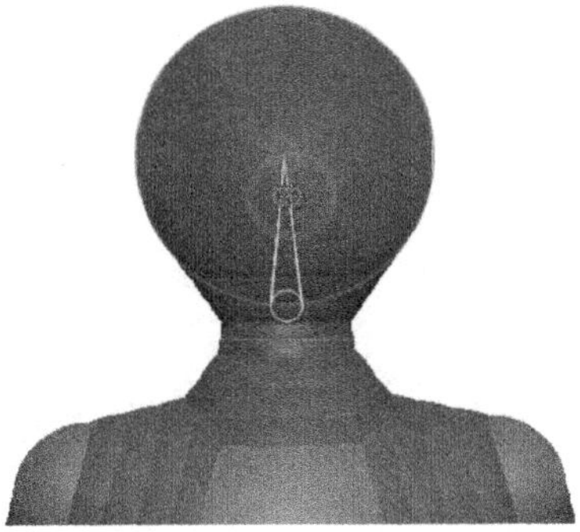

Figure 5-26 *ZSphere created for the lower part of the head*

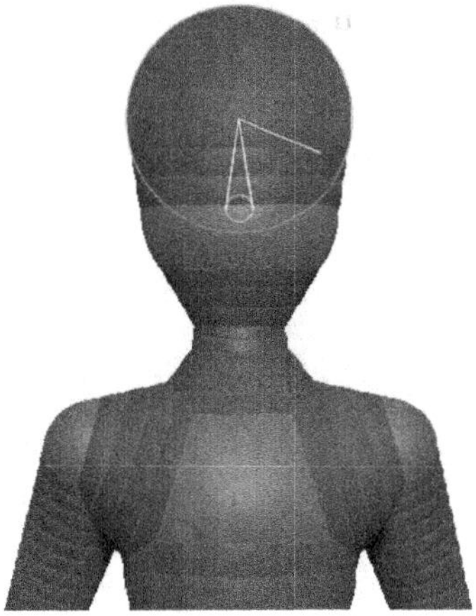

Figure 5-27 *ZSphere created for the upper part of the head*

If required, scale and move the ZSpheres in the structure to adjust the anatomy of the model, refer to Figure 5-28. You can view the skinned preview of the model by pressing the A key, refer to Figure 5-29.

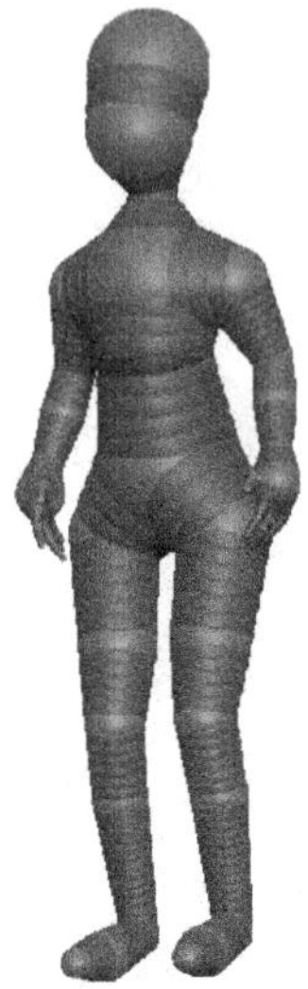

Figure 5-28 *Final model of the body*

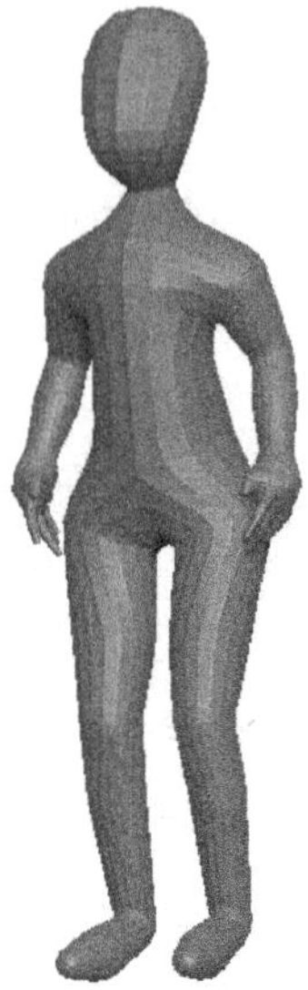

Figure 5-29 *Skinned preview of the body*

Saving the Model

In this section, you will save the file using the steps given next.

1. Choose **Save As** button from the **Tool** palette, the **Save ZTool** dialog box is displayed. In this dialog box, browse to the location *\Documents\ZBrushprojects\c05.*

2. Enter **c05tut1** in the **File name** edit box and then choose the **Save** button.

EXERCISES

The output of the models used in the following exercises can be accessed by downloading the *c05_ZBrush_2018_exr.zip* file from *www.cadsofttech.com*. The path of the file is as follows: *Textbooks > Animation and Visual Effects > ZBrush > Pixologic ZBrush 2018 for Novices.*

Exercise 1

Create the model of a tree using ZSpheres, as shown in Figure 5-30.

(Expected time: 30 min)

Figure 5-30 *Model of a tree*

Exercise 2

Create the scene shown in Figure 5-31 using ZSpheres and different subtools.

(Expected time: 30 min)

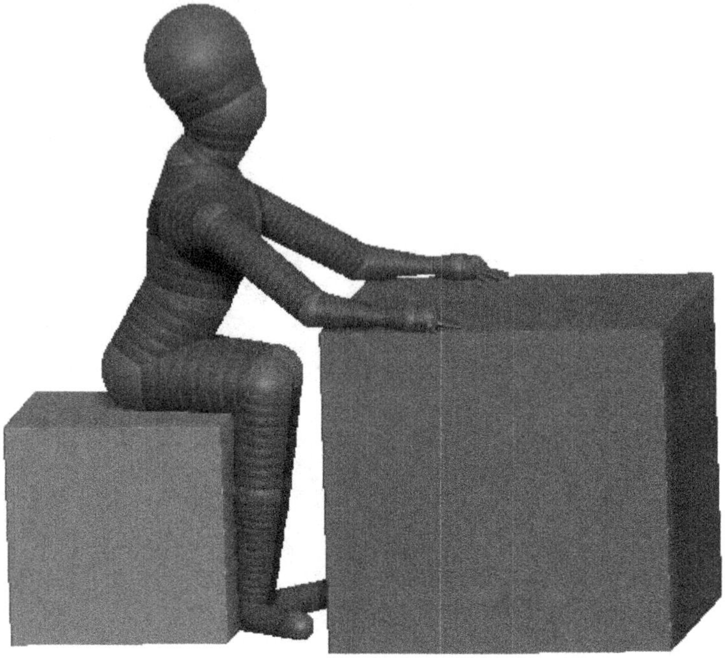

Figure 5-31 *Scene to be created for Exercise 2*

Chapter 6

DynaMesh, NanoMesh, and ZRemesher

Learning Objectives

After completing this chapter, you will be able to:

- *Work with DynaMesh*
- *Work with ZRemesher*
- *Create different models using the DynaMesh tool*

INTRODUCTION

In ZBrush, you can create various types of models using different techniques. The models created in ZBrush have extremely high level of details that cannot be achieved in other modeling software applications. ZBrush uses brush based approach to add details to a model. Different studios such as Luma Pictures, Id Software, and Meteor Studio have used ZBrush for character creation. For instance, the characters Davy Jones in the movie Pirates of the Caribbean and Na'vi in the movie Avatar were created with ZBrush. In this chapter, you will learn about Dynamesh and ZRemesher that help you in creating high detail models.

TUTORIALS

All the files used in the tutorial can be downloaded from the CADSoft website (*www.cadsofttech.com*). These files are compressed in zip file format and are required to be extracted before using them in the tutorials. The path of the files is as follows: *Textbooks > Animation and Visual Effects > Pixologic ZBrush 2018 > Pixologic ZBrush 2018 for Novices*

Tutorial 1

In this tutorial, you will create a coffee mug using a cylinder and the **DynaMesh** tool. The final output of the model is shown in Figure 6-1. **(Expected time: 20 min)**

Figure 6-1 *A coffee mug*

The following steps are required to complete this tutorial:

a. Create the body of the mug.
b. Create the handle of the mug.
c. Save the model.

Creating the Body of the Mug

In this section, you will create the body of the mug using a cylinder and the **DynaMesh** tool.

1. Choose the **Init ZBrush** button from the **Preferences** palette; the message box is displayed. Choose the **Yes** button from the message box; ZBrush is initialized to its default state.

2. Choose the Current Tool button from the **Tool** palette; a flyout is displayed. Choose the **Cylinder3D** primitive from this flyout and create it in the canvas.

3. Choose the **Edit** button from the top shelf.

4. Press and hold the left mouse button and drag the cursor downward in the canvas area such that the cylinder is rotated and stands vertically. While dragging the cursor, press the SHIFT key; the cylinder is snapped at right angle to the canvas.

5. Choose the **Make PolyMesh3D** button from the **Tool** palette to convert the primitive cylinder into a polymesh. Next, choose the **PolyF** button in the right shelf; the polygon edges are displayed in the cylinder, refer to Figure 6-2.

6. Expand the **Geometry** subpalette in the **Tool** palette. In this subpalette, expand the **DynaMesh** option and then choose the **DynaMesh** button; the polygon distribution in the cylinder will be modified, refer to Figure 6-3.

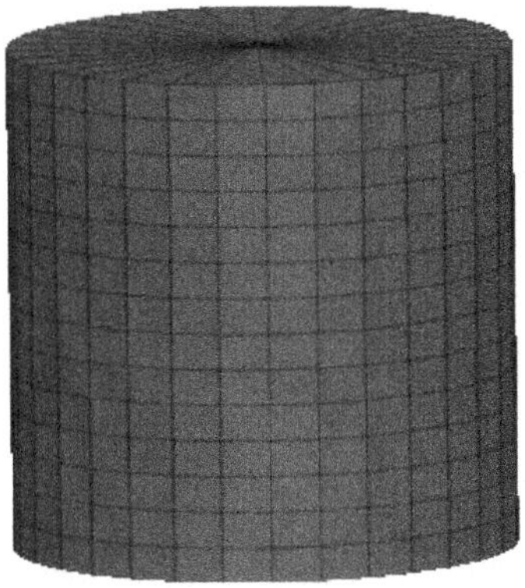

Figure 6-2 *The polygon edges displayed on choosing the **PolyF** button*

Figure 6-3 *The polygon distribution modified*

7. Press and hold the left mouse button and drag the cursor downward in such a way that the cylinder is rotated and its top is displayed. While dragging the cursor, press the SHIFT key; the cylinder is snapped to the canvas.

8. Expand the **SubTool** subpalette in the **Tool** palette. Next, in this subpalette, choose the **Rename** button; the **Please enter subtool title** window consisting of a text box is displayed. Enter **coffee mug** in this text box; the cylinder primitive is renamed as *coffee mug*.

9. Choose the Current Brush button from the left shelf; a flyout containing different sculpting brushes is displayed. Choose the **MaskCircle** brush from this flyout; a message box asking you to press CTRL to use this brush will be displayed. Choose the **OK** button to close this message box.

10. Press and hold the CTRL key and choose the Current Stroke button; a flyout is displayed. In this flyout, choose the **Square** button. Similarly, choose the **Center** button located at the bottom of the flyout, refer to Figure 6-4. Next, release the CTRL key.

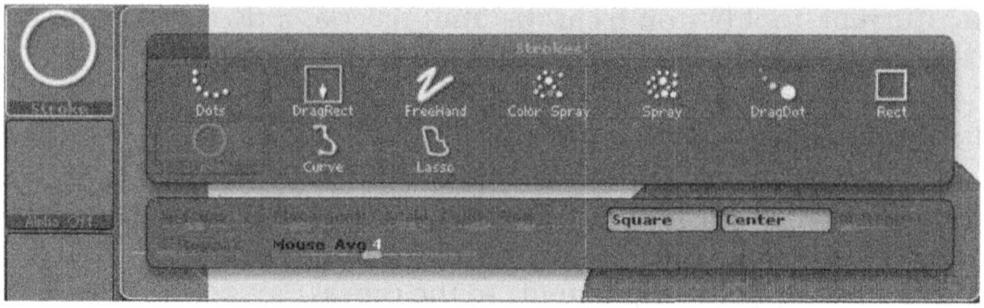

***Figure 6-4** The **Square** and **Center** buttons chosen from the flyout*

The **Square** and **Center** buttons will enable you to start the brush stroke from the centre of the *coffee mug* and create a perfect circle.

11. Activate the symmetry in the X-axis by pressing the X key.

12. Press and hold the CTRL key and place the cursor at the position where two red dots representing the symmetry overlap. Next, drag the cursor on the surface of *coffee mug* to create a circular stroke on the surface of the *coffee mug*, refer to Figure 6-5; a circular mask is applied on the top of the *coffee mug*.

    On rotating the view of the *coffee mug*, you will notice that the circular mask is applied on the bottom of the *coffee mug* also.

13. Choose the Current Brush button from the left shelf; a flyout containing different sculpting brushes is displayed. Choose the **MaskRect** brush from this flyout; a message box asking you to press CTRL to use this brush will be displayed. Choose the **OK** button to close this message box.

14. Press and hold the CTRL+ALT keys and drag the cursor at the bottom of the *coffee mug* to create a rectangular stroke, refer to Figure 6-6; the mask is removed from the bottom area.

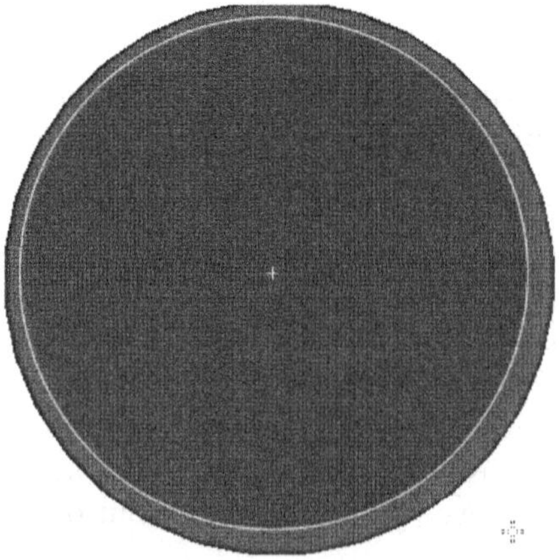

***Figure 6-5** A circular stroke created on the top of the coffee mug*

***Figure 6-6** A rectangular stroke created at the bottom of the coffee mug*

You can remove a mask from a particular area by pressing CTRL+ALT, and then dragging the cursor on that particular area.

15. Choose the **PolyF** button from the **Transform** palette; the polygon edges disappear from the surface of the *coffee mug*, refer to Figure 6-7.

16. Press CTRL+I; the mask is inverted, refer to Figure 6-8.

Figure 6-7 *Polygon edges disappeared from the surface*

Figure 6-8 *The mask inverted on the surface*

17. Expand the **Deformation** subpalette in the **Tool** palette.

You need to change the offset of the unmasked area of the *coffee mug* along the Z-axis using the **Offset** slider.

18. Deactivate the x option corresponding to the **Offset** slider by choosing the **x** button located on the right side of the slider. Next, choose the **z** button to activate the z option.

19. Drag the **Offset** slider toward left; the unmasked area moves downward, refer to Figure 6-9. Continue dragging the slider till the unmasked area reaches the bottom of the *coffee mug*.

20. Press and hold the CTRL key and drag the cursor on the canvas area; the mask is removed from the surface of the *coffee mug*. You will notice that the geometry is distorted, refer to Figure 6-10.

Figure 6-9 *The unmasked area moved downward*

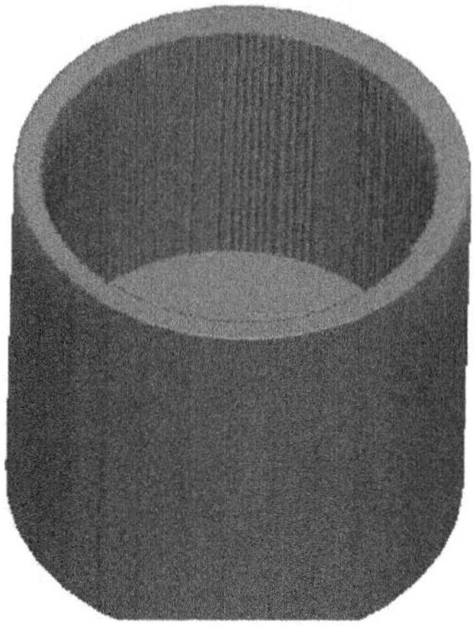

Figure 6-10 *The distorted geometry of the coffee mug*

21. Press and hold the CTRL key and drag the cursor on the canvas area again; some distortions are removed from the surface of the *coffee mug*. To remove the distortions completely, press and hold the SHIFT key; the **Smooth** brush is activated. Next, drag the cursor on the surface of the *coffee mug*; the surface of the *coffee mug* is smoothened, refer to Figure 6-11.

22. Make sure the **Activate Symmetry** button in the **Transform** palette is chosen. Next, choose the **(R)** button; the **RadialCount** slider is activated. By default, the **>X<** button is chosen. To deactivate this button, choose it again. Next, choose the **>Z<** button; the radial symmetry in the Z-axis is activated. In the **RadialCount** edit box, enter the value **73**, refer to Figure 6-12. This edit box is displayed when you click on the **RadialCount** slider.

Figure 6-11 *The surface of the coffee mug smoothened*

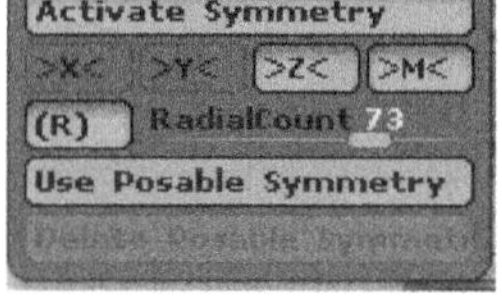

Figure 6-12 *The value of the **RadialCount** slider set to **73***

23. Choose the Current Brush button from the left shelf; a flyout containing different sculpting brushes is displayed. Choose the **Move** brush from this flyout. Set the value of the **Draw Size** slider to **120**.

24. Press and hold the left mouse button, and drag the lower part of the *coffee mug* inward using the **Move** brush, as shown in Figure 6-13.

25. Choose the Current Brush button from the left shelf; a flyout containing different sculpting brushes is displayed. Choose the **Standard** brush from this flyout.

26. Choose the **Zsub** button from the top shelf and set the value of the **Z Intensity** slider to **30**.

27. Set the value of the **Draw Size** slider to **10**, and create a pattern, as shown in Figure 6-14.

Figure 6-13 *The lower part of the coffee mug moved inward*

Figure 6-14 *A pattern created on the coffee mug*

28. Create the same pattern below the pattern created in Step 27, refer to Figure 6-15. Next, deactivate the symmetry by choosing the **Activate Symmetry** button in the **Transform** palette.

Creating the Handle of the Mug

In this section, you will create the handle of the mug using the **InsertHRing** brush.

1. Choose **Insert-IMM** from the **Brush** tab of the **LightBox** browser located at the top of the canvas by double-clicking on it; the list of brushes is displayed. Now, choose the **InsertHRing** brush from the list and drag the cursor on the right side of the *coffee mug* to create the handle of the *coffee mug*; a handle is created and a mask is applied on the *coffee mug*, refer to Figure 6-16.

2. Choose the Current Brush button from the left shelf; a flyout containing different sculpting brushes is displayed. Choose the **Move** brush from this flyout and refine the shape of the handle using the **Move** brush, refer to Figure 6-17.

3. Press and hold the CTRL key and drag the cursor on the canvas area; the mask is removed from the surface of the *coffee mug*.

4. Press and hold the CTRL key and drag the cursor on the canvas area again; the handle is merged with the *coffee mug* and the final model is created, refer to Figure 6-18.

Figure 6-15 *The same pattern created on the coffee mug*

Figure 6-16 *A handle created using the* ***InsertHRing*** *brush*

Figure 6-17 *The shape of the handle modified using the* ***Move*** *brush*

Figure 6-18 *The final model of the coffee mug*

Saving the Model

In this section, you will save the file using the steps given next.

1. Choose **Save As** button from the **Tool** palette, the **Save ZTool** dialog box is displayed. In this dialog box, browse to the location *\Documents\ZBrushprojects\c06*.

2. Enter **c06tut1** in the **File name** edit box and then choose the **Save** button.

Tutorial 2

In this tutorial, you will create a cartoon octopus using a sphere and the **DynaMesh** tool. The final output of the model is shown in Figure 6-19. **(Expected time: 20 min)**

Figure 6-19 *The final model of the cartoon octopus*

The following steps are required to complete this tutorial:

a. Create the body of the octopus.
b. Create the tentacles of the octopus.
c. Create the eyes of the octopus.
d. Add detail to the octopus.
e. Save the model.

Creating the Body of the Octopus

In this section, you will create the body of the octopus using a sphere and the **DynaMesh** tool.

1. Choose the **Init ZBrush** button from the **Preferences** palette; the message box is displayed. Choose the **Yes** button from the message box; ZBrush is initialized to its default state.

2. Choose the Current Tool button from the **Tool** palette; a flyout is displayed. Choose the **Sphere3D** primitive from this flyout and create it in the canvas, refer to Figure 6-20.

3. Choose the **Edit** button from the top shelf. Next, choose the **PolyF** button from the **Transform** palette; the polygon edges on the sphere are displayed.

4. Press and hold the left mouse button and drag the cursor downward in the canvas area such that the sphere is rotated. While dragging the cursor, press the SHIFT key; the sphere is snapped at right angle to the canvas, as shown in Figure 6-21.

***Figure 6-20** The sphere created in the canvas*

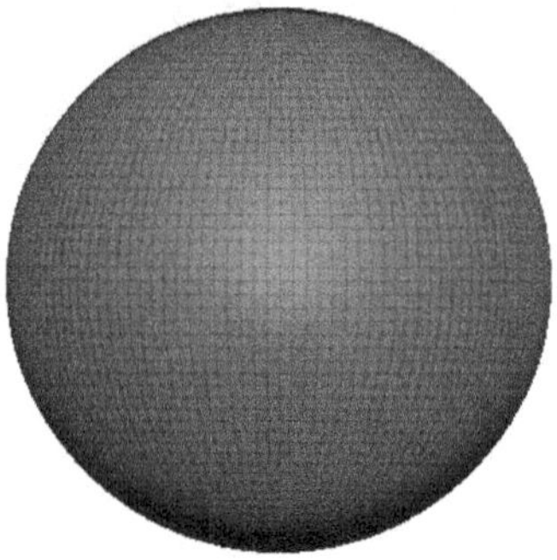

***Figure 6-21** The sphere rotated in the canvas*

5. Choose the Current Brush button from the left shelf; a flyout containing different sculpting brushes is displayed. Choose the **ClipRect** brush from this flyout; a message box is displayed, asking you to press CTRL+SHIFT keys to activate this brush. Choose the **OK** button to close this message box.

6. Press CTRL+SHIFT keys and then press and hold the left mouse button. Next, hover the cursor over the upper part of the sphere; a rectangular marquee selection appears. Select the upper part of the sphere, refer to Figure 6-22. Now, release the left mouse button and CTRL+SHIFT keys; the lower part of the sphere is deleted, refer to Figure 6-23.

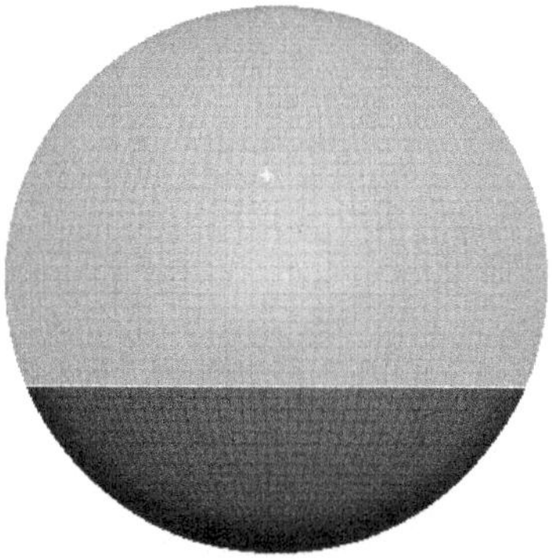

***Figure 6-22** The rectangular marquee selection created on the surface of sphere*

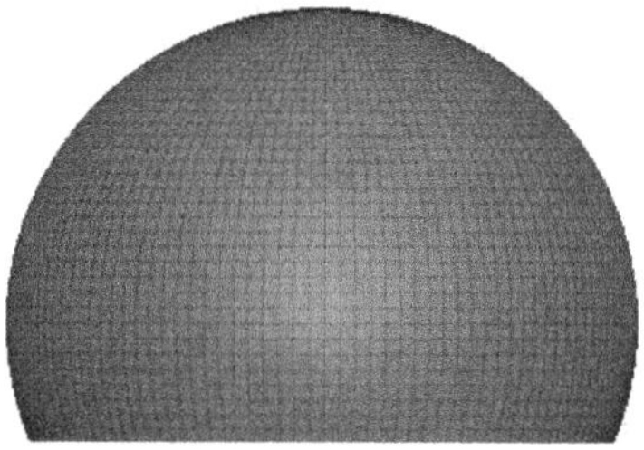

***Figure 6-23** The lower part of the sphere deleted*

7. Choose the **Make PolyMesh3D** button from the **Tool** palette to convert the primitive sphere into polymesh. Next, choose the **PolyF** button from the right shelf; the polygon edges on the sphere disappear.

8. Expand the **Geometry** subpalette in the **Tool** palette. In this subpalette, choose the **DynaMesh** button in the **DynaMesh** area; the sphere becomes smoother.

9. Expand the **SubTool** subpalette in the **Tool** palette. Next, in this subpalette, choose the **Rename** button; the **Please enter subtool title** window consisting of a text box is displayed. Enter **octopus** in this text box; the sphere primitive is renamed as *octopus*.

10. Choose the **Activate Symmetry** button from the **Transform** palette. Next, choose the **(R)** button; the **RadialCount** slider is activated. By default, the **>X<** button is chosen. To deactivate this button, choose it again. Next, choose the **>Z<** button; the radial symmetry in the Z-axis is activated. Set the value of the **RadialCount** slider to **50**, refer to Figure 6-24.

11. Choose the Current Brush button from the left shelf; a flyout containing different sculpting brushes is displayed. Choose the **Move** brush from this flyout and create the shape of the body using the radial symmetry and the **Move** brush, refer to Figure 6-25. If required, increase the brush size to create the shape of the body.

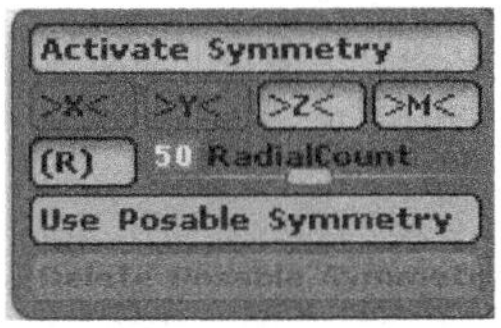

Figure 6-24** The value of the **RadialCount** slider set to **50

***Figure 6-25** The shape modified using the **Move** brush*

Creating the Tentacles of the Octopus

In this section, you will create the tentacles of the octopus using the **Inflat** and **Move** brushes.

1. Choose the **Transform** palette. In this palette, set the value of the **RadialCount** slider to **8**, refer to Figure 6-26.

2. Choose the Current Brush button from the left shelf; a flyout containing different sculpting brushes is displayed. Choose the **Inflat** brush from this flyout.

3. Decrease the brush size and drag the cursor on the lower part of the *octopus* to create base for the tentacles, refer to Figure 6-27. In this figure, you will notice that the geometry is stretched. To remove the stretching, press and hold the CTRL key and drag the cursor on the canvas area; the *octopus* is smoothened, refer to Figure 6-28.

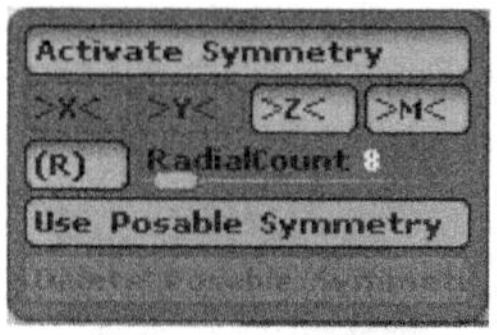

Figure 6-26 *The value of the* ***RadialCount*** *slider set to* ***8***

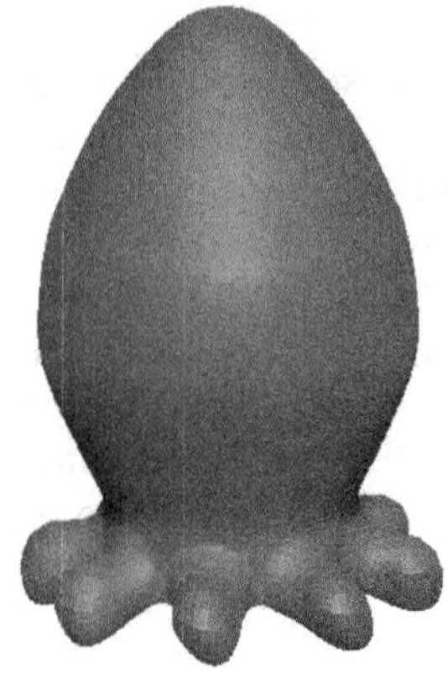

Figure 6-27 *Base for the tentacles created using the* ***Inflat*** *brush*

4. Choose the Current Brush button from the left shelf; a flyout containing different sculpting brushes is displayed. Choose the **Move** brush from this flyout and drag out the tentacles using this brush, refer to Figure 6-29.

Figure 6-28 *The surface of the octopus smoothened*

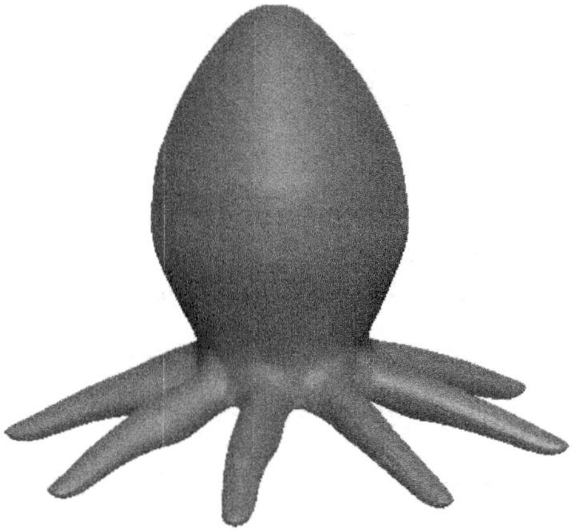

Figure 6-29 *The tentacles dragged out using the* ***Move*** *brush*

5. Drag the ends of the tentacles toward left using the **Move** brush, refer to Figure 6-30. The geometry is stretched again. To remove the stretching, press and hold the CTRL key and drag the cursor on the canvas area.

6. Press and hold the SHIFT key; the **Smooth** brush is activated. Next, with the SHIFT key pressed, drag the cursor at the ends of the tentacles to make them pointed and smooth, refer to Figure 6-31.

Figure 6-30 Ends of the tentacles moved toward left

*Figure 6-31 Ends of the tentacles smoothened using the **Smooth** brush*

7. Choose the Current Brush button from the left shelf; a flyout containing different sculpting brushes is displayed. Choose the **SnakeHook** brush from this flyout and drag out the tentacles toward left using this brush, refer to Figure 6-32.

8. Choose the Current Brush button from the left shelf; a flyout containing different sculpting brushes is displayed. Choose the **Inflat** brush from this flyout and drag the cursor on the surface of the tentacles to add depth to the tentacles, refer to Figure 6-33. If required, use the **Smooth** brush on the ends of the tentacles to make them pointed.

*Figure 6-32 Ends of the tentacles moved toward left using the **SnakeHook** brush*

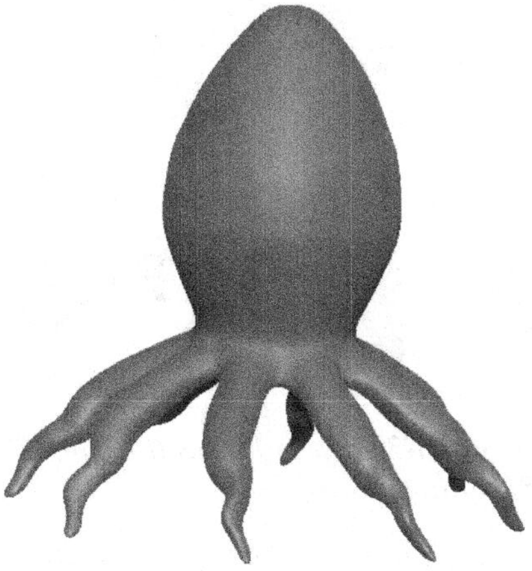

*Figure 6-33 Depth added to the tentacles using the **I3flat** brush*

Creating the Eyes of the Octopus

In this section, you will create the eyes of the octopus using the **InsertSphere** brush.

1. Choose the **(R)** and **>Z<** buttons in the **Transform** palette; the radial symmetry in the Z-axis is deactivated.

2. Press the X key to activate symmetry in the X-axis.

3. Choose the Current Brush button from the left shelf; a flyout containing different sculpting brushes is displayed. Choose the **Standard** brush from this flyout and then choose the **Zsub** button from the top shelf.

4. Decrease the brush size, if required. Next, drag the cursor on the surface of the *octopus* to create an eye socket, refer to Figure 6-34.

5. Choose the Current Brush button from the left shelf; a flyout containing different sculpting brushes is displayed. Choose the **Zadd** button and then choose the **InsertSphere** brush from this flyout and drag on the surface of eye sockets to create eyeball; a mask is applied on the surface of *octopus,* refer to Figure 6-35.

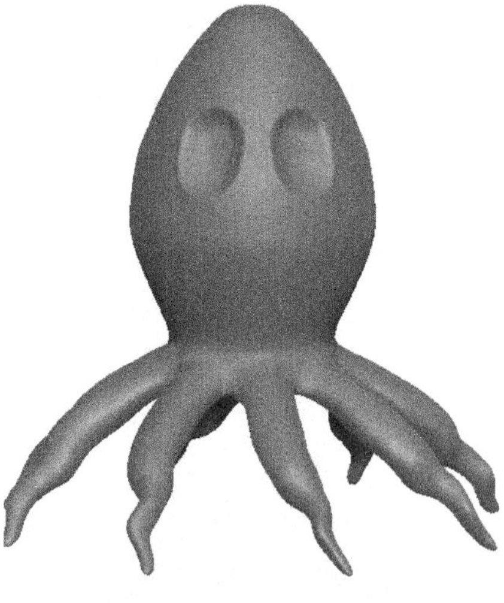

Figure 6-34 *Eye socket created using the* ***Standard*** *brush*

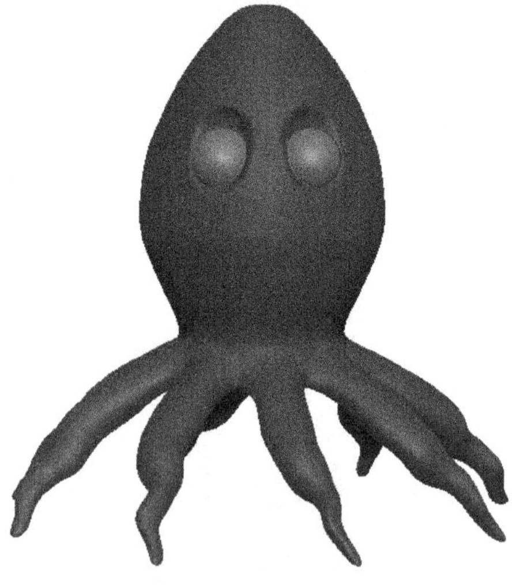

Figure 6-35 *Eye ball created using the* ***InsertSphere*** *brush*

6. Press and hold the CTRL key and drag the cursor on the canvas area twice; the mask is removed from the surface of *octopus* and the eyeball is merged with the socket, refer to Figure 6-36.

7. Make sure the **InsertSphere** brush is chosen. Press the ALT key and create another sphere on the eyeball, as shown in Figure 6-37. You will notice that a mask is applied on the surface

of *octopus*. To remove the mask from its surface, press and hold the CTRL key and drag the cursor on the open area of the canvas. Next, choose the **Sub** button; a hole is created on the surface of the eye ball.

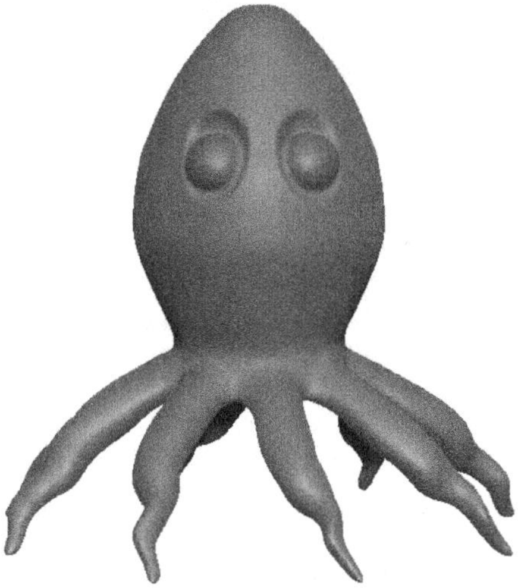

Figure 6-36 *Eyeball merged with the socket*

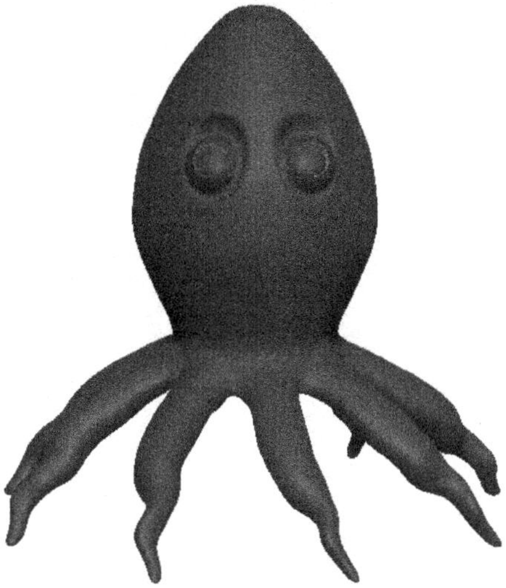

Figure 6-37 *A sphere created on eyeball with the ALT key pressed*

8. Create another smaller sphere on the hole created in Step 7 using the **InsertSphere** brush, refer to Figure 6-88. Press and hold the CTRL key and drag the cursor on the canvas area twice; the mask is removed from the surface of *octopus* and the newly created sphere is merged with the socket.

Adding Detail to the Octopus

In this section, you will add detail to the octopus using different alpha images.

1. Deactivate the symmetry in the X-axis by pressing the X key. Choose the Current Brush button from the left shelf; a flyout containing different sculpting brushes is displayed. Choose the **Standard** brush from this flyout.

2. Choose the Current Stroke button from the left shelf; a flyout containing different types of strokes is displayed. Choose the **DragRect** stroke from this flyout.

3. Choose the Current Alpha button from the left shelf; a flyout containing different alpha images is displayed. Choose the **Alpha 17** alpha image from this flyout and create a pattern on one of the tentacles of the *octopus*, refer to Figure 6-39. Continue creating the same pattern on all the tentacles, refer to Figure 6-40.

Figure 6-38 *Sphere created on the surface of the hole*

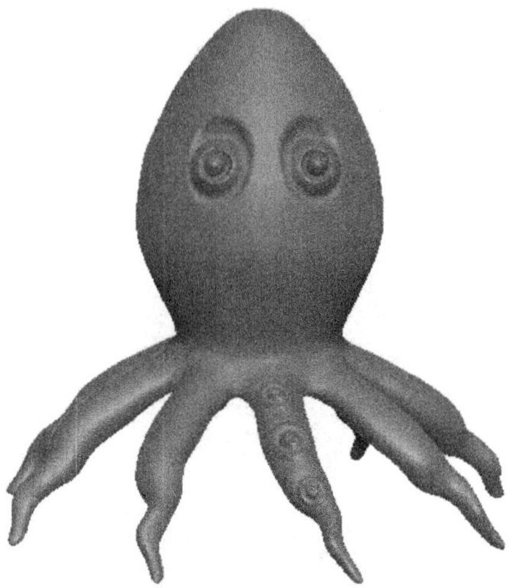

Figure 6-39 *Pattern created on a tentacle*

4. Choose the Current Stroke button from the left shelf; a flyout containing different types of strokes is displayed. Choose the **FreeHand** stroke from this flyout.

5. Choose the Current Alpha button from the left shelf; a flyout containing different alpha images is displayed. Choose the **Alpha 24** alpha image from this flyout. Next, increase the brush size and set the value of the **Z Intensity** slider in the top shelf to **4**. Press and hold the left mouse button and drag the cursor on the body of the *octopus* to create a noise pattern, as shown in Figure 6-41.

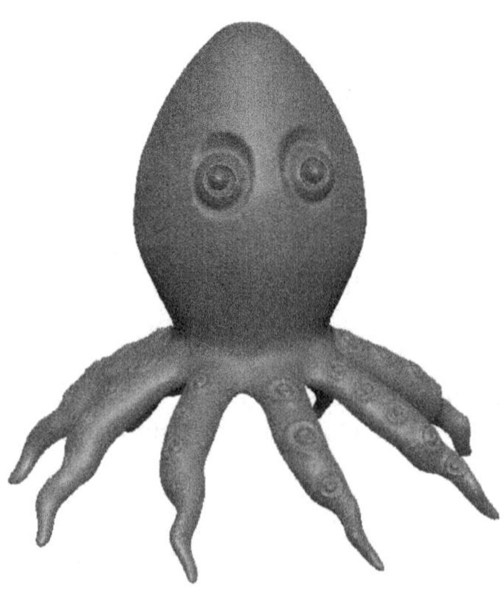

Figure 6-40 *Pattern created on all the tentacles*

Figure 6-41 *Noise pattern created on the body of the octopus*

Saving the Model

In this section, you will save the file using the steps given next.

1. Choose **Save As** button from the **Tool** palette, the **Save ZTool** dialog box is displayed. In this dialog box, browse to the location *\Documents\ZBrushprojects\c06*.

2. Enter **c06tut2** in the **File name** edit box and then choose the **Save** button.

EXERCISE

The output of the model used in the exercise can be accessed by downloading the *c06_ZBrush_2018_exr.zip* file from *www.cadsofttech.com*. The path of the file is as follows: *Textbooks > Animation and Visual Effects > ZBrush > Pixologic ZBrush 2018 for Novices*.

Exercise 1

Create the model of a cartoon monkey using the **DefaultWaxSphere.ZPR** file and the DynaMesh tool, as shown in Figure 6-42. **(Expected time: 30 min)**

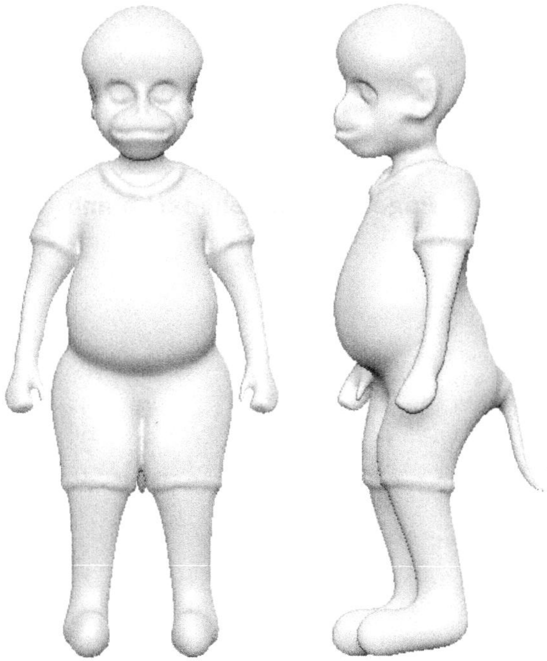

Figure 6-42 *Model of the cartoon monkey*

Page intentionally left blank

Chapter 7

ShadowBox

Learning Objectives

After completing this chapter, you will be able to:

- *Work with the ShadowBox*
- *Create different models using the ShadowBox*

INTRODUCTION

In this chapter, you will learn to create different models using the ShadowBox tool. This tool is used to create basic shapes and not the highly detailed models. However, the basic models created in ShadowBox can be sculpted later using different brushes.

TUTORIAL

All the files used in the tutorial can be downloaded from the CADSoft website (*www.cadsofttech.com*). These files are compressed in zip file format and are required to be extracted before using them in the tutorials. The path of the files is as follows: *Textbooks > Animation and Visual Effects > Pixologic ZBrush 2018 > Pixologic ZBrush 2018 for Novices*

Tutorial 1

In this tutorial, you will create the model of a key using the ShadowBox. The final output of the model is shown in Figure 7-1. **(Expected time: 20 min)**

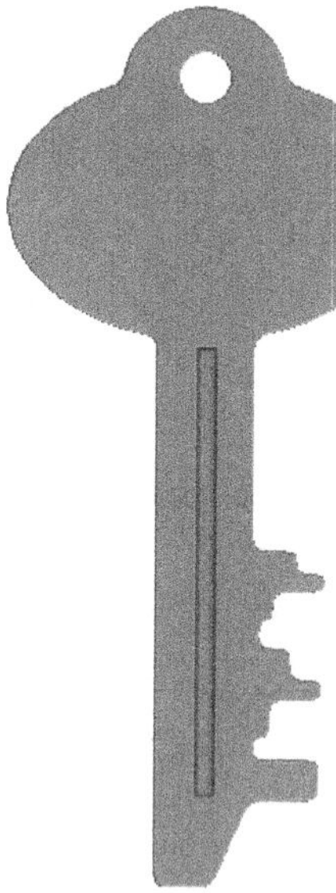

Figure 7-1 *The model of a key*

The following steps are required to complete this tutorial:

a. Create the shape of the key in the ShadowBox.
b. Modify the key outside the ShadowBox.
c. Save the model.

Creating the Shape of the Key in the ShadowBox

In this section, you will create the shape of the key using the Mask brushes on the planes of ShadowBox.

1. Choose the **Init ZBrush** button from the **Preferences** palette; the message box is displayed. Choose the **Yes** button from the message box; ZBrush is initialized to its default state.

2. Choose the Current Tool button from the **Tool** palette; a flyout is displayed. Choose the **Sphere3D** primitive from this flyout and create it in the canvas.

3. Choose the **Edit** button from the top shelf.

4. Choose the **Make PolyMesh3D** button from the **Tool** palette to convert the primitive sphere into a polymesh.

5. Expand the **Geometry** subpalette in the **Tool** palette. In this subpalette, set the value of **Res** slider in the **ShadowBox** area to **400**, refer to Figure 7-2.

6. Choose the **ShadowBox** button in the ShadowBox area; a ShadowBox containing a 3D mesh is displayed in the canvas. Rotate the canvas such that you can view the front view of the ShadowBox, refer to Figure 7-3.

Figure 7-2 *The value of the* ***Res*** *slider set to* ***400***

Figure 7-3 *The front view of the ShadowBox displayed in the canvas*

7. Press and hold the CTRL key and then drag the cursor in the canvas area; the 3D mesh disappears from the ShadowBox. Rotate the canvas such that you can view the back side of the front plane.

8. Choose the Current Brush button from the left shelf; a flyout containing different sculpting brushes is displayed. Choose the **MaskCircle** brush from this flyout; a message box asking you to press CTRL to use this brush is displayed. Choose the **OK** button to close this message box.

9. Press and hold the CTRL key and choose the Current Stroke button; a flyout is displayed. In this flyout, choose the **Center** button located at the bottom of the flyout, refer to Figure 7-4. Next, release the CTRL key.

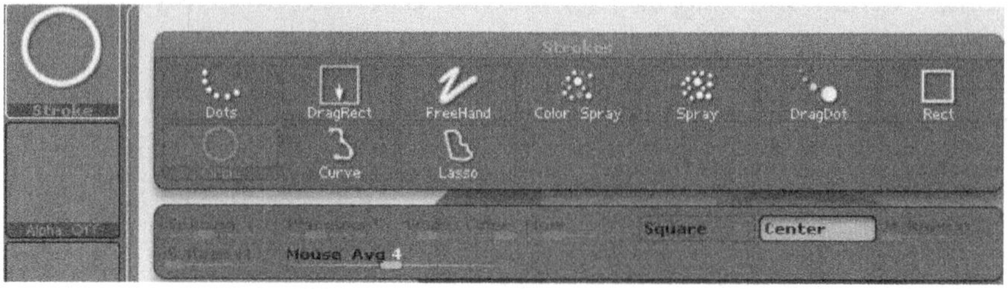

***Figure 7-4** The **Center** button chosen from the flyout*

10. Press and hold the CTRL key and drag the cursor on the front plane of the ShadowBox to create an elliptical shape on the plane, refer to Figure 7-5. Release the CTRL key; a mask is created in the front plane.

11. Press and hold the CTRL key and choose the Current Stroke button; a flyout is displayed. In this flyout, choose the **Square** button located at the bottom of the flyout. This button enables you to create a perfect circle. Next, release the CTRL key.

12. Press and hold the CTRL key and drag the cursor on the top of the mask created earlier to create a circular stroke, refer to Figure 7-6. Release the CTRL key; another mask is created on the front plane.

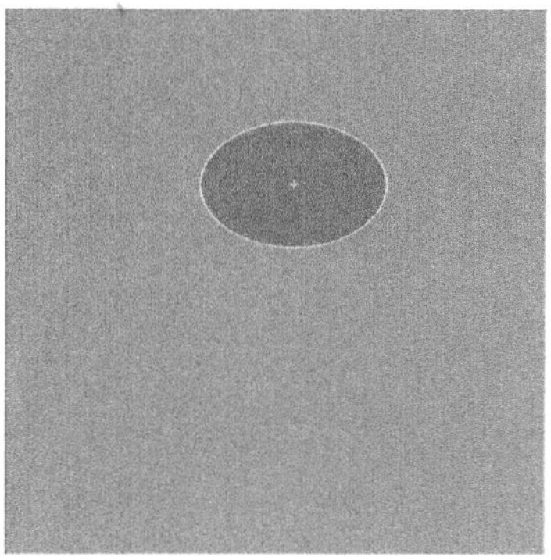

***Figure 7-5** An elliptical stroke created in the backside of the front plane*

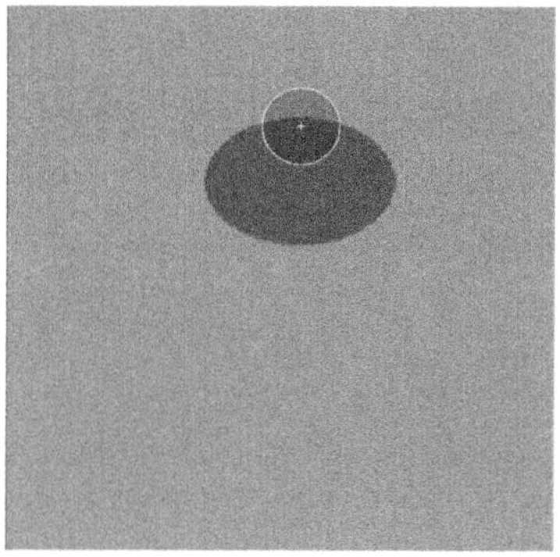

***Figure 7-6** A circular stroke created above the elliptical mask*

13. Press and hold the CTRL+ALT keys and drag the cursor to create a circular stroke at that location, as shown in Figure 7-7. Release the CTRL key; a hole is created in the mask and the 3D object, refer to Figure 7-8.

14. Choose the Current Brush button from the left shelf; a flyout containing different sculpting brushes is displayed. Choose the **MaskRect** brush from this flyout; a message box asking you to press CTRL to use this brush is displayed. Choose the **OK** button to close this message box.

15. Press and hold the CTRL key and drag the cursor on the front plane of the ShadowBox to create a rectangular stroke below the mask created earlier. Release the CTRL key; a mask is created in the front plane, refer to Figure 7-9.

16. Press and hold the CTRL key and drag the cursor in the front plane to create another rectangular stroke on the right side of the mask created earlier. Release the CTRL key; another rectangular mask is created in the front plane, refer to Figure 7-10.

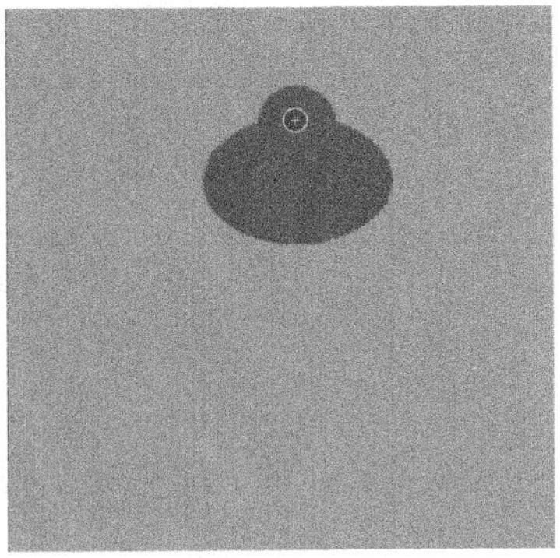

Figure 7-7 *A circular stroke created on the mask with the ALT key pressed*

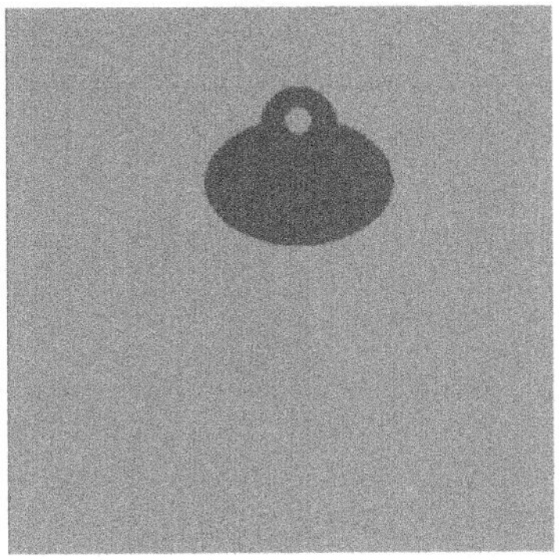

Figure 7-8 *A hole created in the mask*

Figure 7-9 *A rectangular mask created using the **MaskRect** brush*

Figure 7-10 *A rectangular mask created on the right side of the mask*

17. Press and hold the CTRL+ALT keys and drag the cursor to create another rectangular stroke on the top of the rectangular mask created earlier, refer to Figure 7-11. Release the CTRL+ALT keys; the area where the stroke was created gets deleted and the shape of the mask is modified, refer to Figure 7-12.

18. Press and hold the CTRL key and drag the cursor to create a rectangular stroke below the mask created in Step 15. Release the CTRL key; a mask is created, refer to Figure 7-13.

19. Similarly, create more rectangular strokes to create the bottom part of the key using the **MaskRect** brush, refer to Figure 7-14.

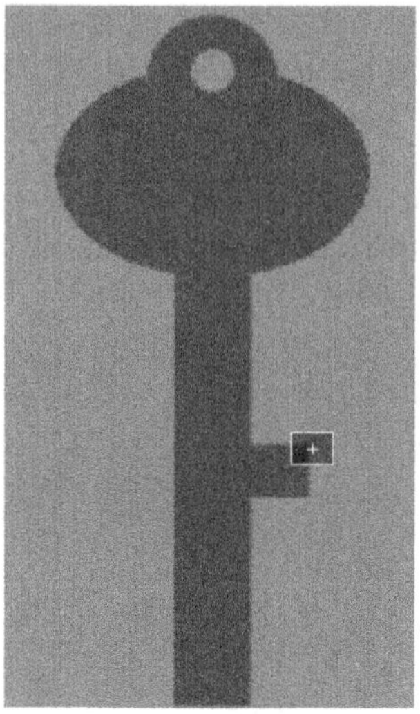

Figure 7-11 *A rectangular stroke created on the mask with ALT key pressed*

Figure 7-12 *The shape of the mask modified*

Figure 7-13 *Shape of the mask modified using the **MaskRect** brush*

Figure 7-14 *Bottom part of the key created using the **MaskRect** brush*

20. Choose the Current Brush button from the left shelf; a flyout containing different sculpting brushes is displayed. Choose the **MaskCurve** brush from this flyout; a message box asking you to press CTRL to use this brush is displayed. Choose the **OK** button to close this message box.

21. Press and hold the CTRL+ALT keys and drag the cursor diagonally in the upward direction, as shown in Figure 7-15; the shape of the mask is modified, as shown in Figure 7-16.

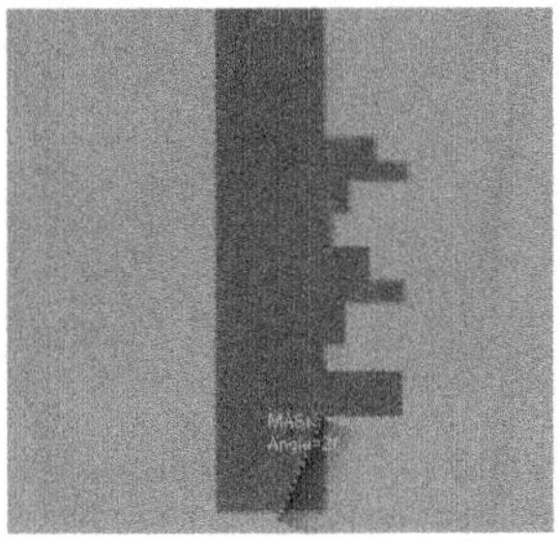

Figure 7-15 *Curved stroke created using the* ***MaskCurve*** *brush*

Figure 7-16 *Shape of the mask modified*

22. Rotate the canvas such that you can view the front view of the ShadowBox; the 3D mesh for the key is displayed in the ShadowBox, refer to Figure 7-17.

23. Rotate the canvas such that you can view the side view of the ShadowBox. On rotating the canvas, you will notice that the depth of the key is too large, refer to Figure 7-18.

Figure 7-17 *A key created in the ShadowBox*

Figure 7-18 *Side view of the key*

24. Rotate the canvas such that you can view the back side view of the side plane of the ShadowBox.

25. Choose the Current Brush button from the left shelf; a flyout containing different sculpting brushes is displayed. Choose the **MaskRect** brush from this flyout; a message box asking you to press CTRL to use this brush is displayed. Choose the **OK** button to close this message box.

26. Press and hold the CTRL key and drag the cursor on the side plane of the ShadowBox to create a rectangular stroke, as shown in Figure 7-19; a rectangular mask is created. Now, rotate the canvas to view the front side of the ShadowBox. On rotating the canvas, you will notice that the depth of the key is decreased, refer to Figure 7-20.

Figure 7-19 *A rectangular stroke created in the side plane*

Figure 7-20 *The depth of the key decreased*

Modifying the Key Outside the ShadowBox

In this section, you will modify the shape of the key outside the ShadowBox.

1. Choose the **Make PolyMesh3D** button from the **Tool** palette, refer to Figure 7-21; the ShadowBox disappears from the canvas and 3D model of the key is displayed, refer to Figure 7-22.

Figure 7-21 *The* ***Make PolyMesh3D*** *button chosen*

Figure 7-22 *The key displayed in the canvas*

2. Choose the Current Brush button from the left shelf; a flyout containing different sculpting brushes is displayed. Choose the **MaskRect** brush from this flyout; a message box asking you to press CTRL to use this brush is displayed. Choose the **OK** button to close this message box.

3. Press and hold the CTRL key and drag the cursor on the lower part of the key to create a rectangular mask, as shown in Figure 7-23.

4. Rotate the canvas such that you can view the back side view of the key. You will notice that a mask is applied on the backside of the key also. You need to remove the mask.

5. Rotate the canvas such that you can view the side view of the key. Press and hold the CTRL+ALT keys and drag the cursor to create a rectangular stroke to select the back side of the key. Release the CTRL+ALT keys; mask is removed from the back side of the key, refer to Figure 7-24.

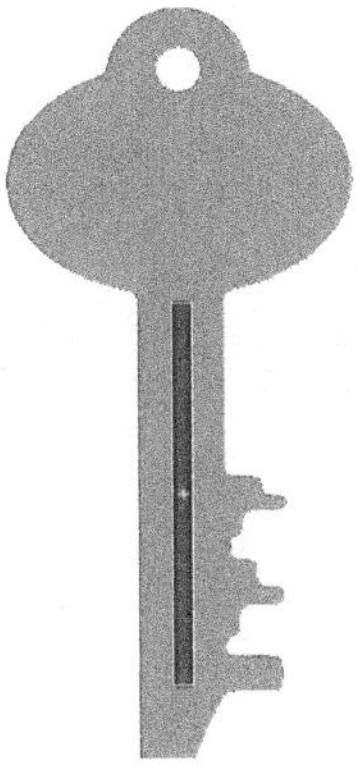

Figure 7-23 *Rectangular stroke created using the* ***MaskRect*** *brush*

Figure 7-24 *Mask removed from the back side of the key*

6. Press CTRL+I; the mask is inverted, refer to Figure 7-25.

7. Expand the **Deformation** subpalette in the **Tool** palette.

   You need to change the offset of the unmasked area of the key along the Y-axis using the **Offset** slider.

8. Deactivate x option corresponding to the **Offset** slider by choosing the **x** button located on the right side of the slider. Next, choose the **y** button to activate the y option, refer to Figure 7-26.

9. Drag the **Offset** slider toward left till the unmasked area moves slightly inward. Press and hold the CTRL key and drag the cursor in the empty area of the canvas; the mask is removed. The final model of the key is created, refer to Figure 7-27.

Figure 7-25 *The mask inverted*

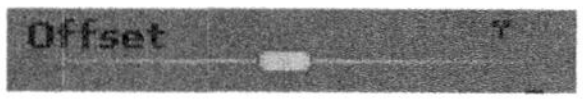

Figure 7-26 *The **Offset** slider*

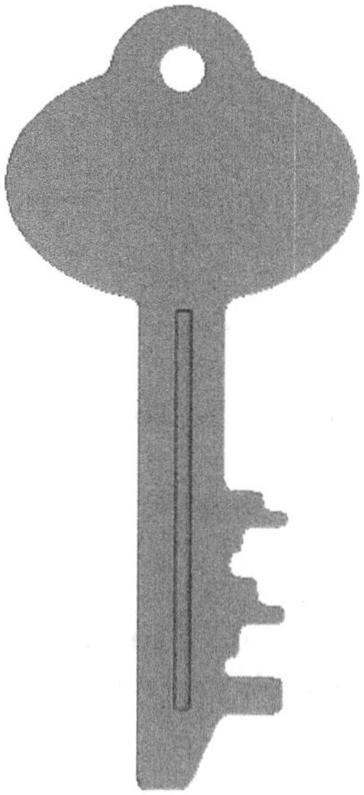

Figure 7-27 *The final model of the key*

Saving the Model

In this section, you will save the file using the steps given next.

1. Choose **Save As** button from the **Tool** palette, the **Save ZTool** dialog box is displayed. In this dialog box, browse to the location *\Documents\ZBrushprojects\c07*.

2. Enter **c07tut1** in the **File name** edit box and then choose the **Save** button.

EXERCISES

The output of the models used in the following exercises can be accessed by downloading the *c07_ZBrush_2018_exr.zip* file from *www.cadsofttech.com*. The path of the file is as follows: *Textbooks > Animation and Visual Effects > ZBrush > Pixologic ZBrush 2018 for Novices*.

Exercise 1

Create the model of a chair using the ShadowBox and Mask brushes, as shown in Figure 7-28.

(Expected time: 20 min)

Figure 7-28 *Model of a chair*

Exercise 2

Create the model of a hut using the ShadowBox and the Mask brushes, as shown in Figure 7-29.

(Expected time: 30 min)

Figure 7-29 *Model of a hut*

Chapter 8

Materials in ZBrush

Learning Objectives

After completing this chapter, you will be able to:

- *Understand the working of the MatCap materials*
- *Understand the working of the Standard materials*
- *Work with the Material palette*
- *Assign different materials to the objects*

INTRODUCTION

In this chapter, you will learn to assign various materials to 3D meshes created in ZBrush. A material is applied to an object to give it a realistic look. ZBrush comes with a library of materials that can either be accessed through the **Material** palette or the left shelf.

TUTORIALS

All the files used in the tutorial can be downloaded from the CADSoft website (*www.cadsofttech.com*). These files are compressed in zip file format and are required to be extracted before using them in the tutorials. The path of the files is as follows: *Textbooks > Animation and Visual Effects > Pixologic ZBrush 2018 > Pixologic ZBrush 2018 for Novices*

Tutorial 1

In this tutorial, you will create the model of an eyeball using different colors. The final output of the model is shown in Figure 8-1. **(Expected time: 20 min)**

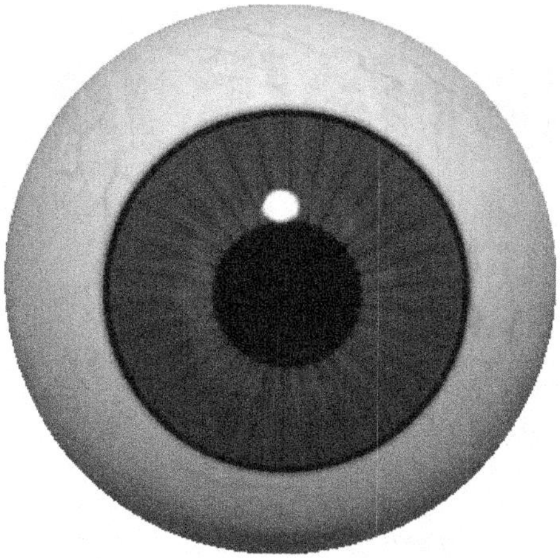

Figure 8-1 *The model of an eyeball*

The following steps are required to complete this tutorial:

a. Create the iris and pupil of the eyeball.
b. Add details to the eyeball.
c. Save the model.

Creating the Iris and Pupil of the Eyeball

In this section, you will create the iris and pupil of the eyeball using different colors.

1. Choose the **Init ZBrush** button from the **Preferences** palette; the message box is displayed. Choose the **Yes** button from the message box; ZBrush is initialized to its default state.

2. Choose the **DynaWaxSphere128.ZPR** file from the **Project** tab of the **LightBox** browser by double-clicking on it; the sphere is created in the canvas.

3. Choose the Current Material button from the left shelf; a flyout containing different materials is displayed. In this flyout, choose the **ToyPlastic** material from the **Standard Materials** area; the material is applied to the sphere. You will notice that the material is gray in color.

4. Choose the white color from the Current Color swatch located below the Current Material button; the white color is applied to the sphere, refer to Figure 8-2.

5. Choose the **Transform** palette. In this palette, make sure the **Activate Symmetry** button is chosen. Next, choose the **(R)** button; the **RadialCount** slider is activated. By default, the **>X<** button is chosen. To deactivate this button, choose it again. Next, choose the **>Z<** button; the radial symmetry in Z-axis is activated. Set the value of the **RadialCount** slider to **82**, refer to Figure 8-3.

Figure 8-2 *The white color applied to the sphere*

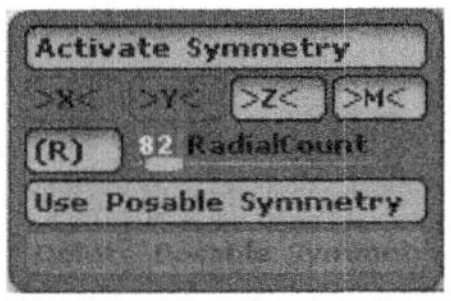

Figure 8-3 *The value of the* ***RadialCount*** *slider set to* ***82***

6. In the **Geometry** subpalette, click on the **Divide** button four times; the value of the **SDiv** slider becomes equal to **5** and the sphere becomes smoother.

7. Choose the **Mrgb** button from the top shelf and then choose the **Zadd** button to deactivate it. Next, choose the **Color** palette. In this palette, choose the **FillObject** button, refer to Figure 8-4.

8. In the **Color** palette, set the values of the **R**, **G**, and **B** sliders to **72**, **46**, and **28**, respectively, as shown in Figure 8-5.

***Figure 8-4** The **FillObject** button chosen in the **Color** palette*

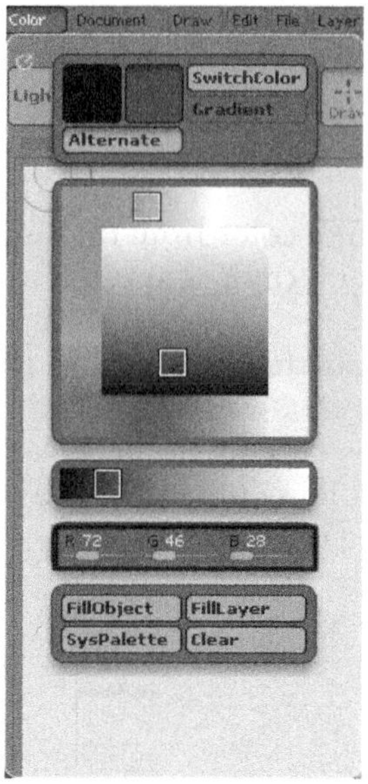

***Figure 8-5** Setting the values of R, G, and B sliders*

9. Hover the cursor on the surface of the sphere, refer to Figure 8-6. Next, drag the cursor inward to create the iris for the eyeball, refer to Figure 8-7.

***Figure 8-6** The cursor hovered on the surface of the sphere*

***Figure 8-7** The iris created for the eyeball*

10. Choose the **Color** palette. In this palette, set the values of the **R**, **G**, and **B** sliders to **5**, **5**, and **5**, respectively. Set the value of the **Draw Size** slider in the top shelf to **3** and hover the cursor at the circumference of the iris, refer to Figure 8-8.

11. Drag the cursor in a circular stroke to create an outline for the iris, refer to Figure 8-9.

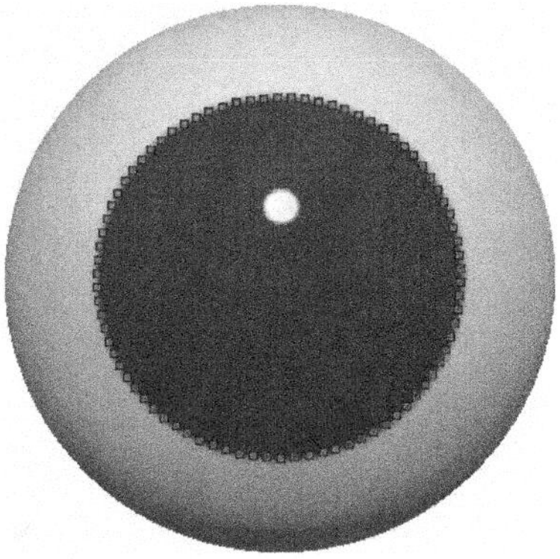

Figure 8-8 The cursor hovered on the circumference of the iris

Figure 8-9 An outline created for the iris

12. Set the value of the **Draw Size** slider in the top shelf to **64.** Hover the cursor at the center of the circle and drag the cursor outward to create a pupil, refer to Figure 8-10.

Adding Details to the Eyeball

In this section, you will add details to the eyeball.

1. Choose the **Transform** palette. In this palette, set the value of the **RadialCount** slider to **30**.

2. Set the value of the **Rgb Intensity** slider in the top shelf to **20** to decrease the intensity of the black color, refer to Figure 8-11. Next, set the value of the **Draw Size** slider to **7**.

Figure 8-10 Pupil created for the eyeball

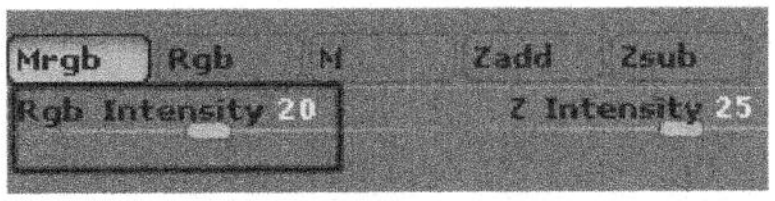

Figure 8-11 The value of the **Rgb Intensity** slider set to **20**

3. Hover the cursor at the circumference of the pupil. Next, drag the cursor outward till it reaches the outline of the iris; thin lines are created on the surface of the iris, refer to Figure 8-12.

4. Choose the **Color** palette. In this palette, set the values of the **R**, **G**, and **B** sliders to **224**, **216**, and **216**, respectively. Next, set the value of the **Rgb Intensity** slider in the top shelf to **3**. Create more thin lines adjacent to the black lines created in Step 3, refer to Figure 8-13.

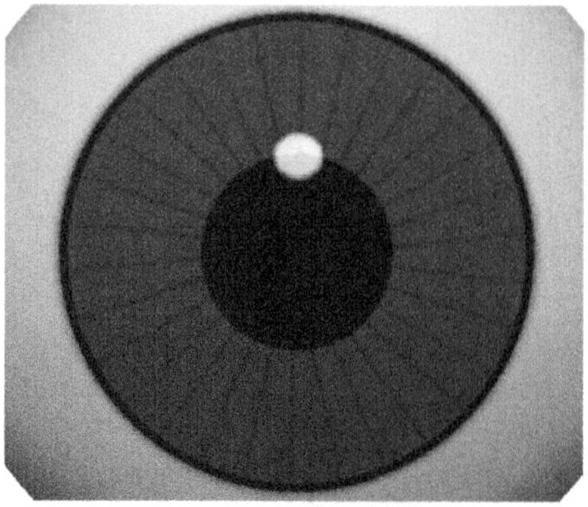

Figure 8-12 Thin lines created on the surface of the iris

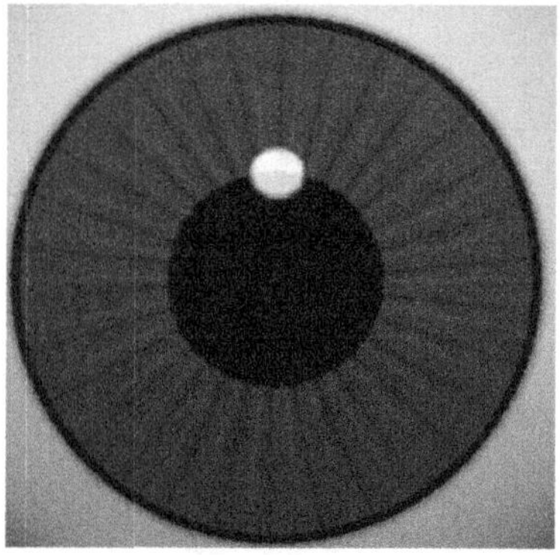

Figure 8-13 More lines created on the surface of the iris

5. Choose the **Transform** palette. In this palette, choose the **Activate Symmetry** button to deactivate symmetry.

6. Choose the **Color** palette. In this palette, set the values of the **R**, **G**, and **B** sliders to **101**, **103**, and **37**, respectively. Set the value of the **Draw Size** slider in the top shelf to **25**. Next, drag the cursor in the area outside the pupil to add more detail to the iris, refer to Figure 8-14.

7. Choose the Current Stroke button from the left shelf; a flyout containing different types of strokes is displayed. Choose the **DragRect** stroke from this flyout.

8. Choose the Current Alpha button from the left shelf; a flyout containing different alpha images is displayed. Choose the **Alpha 22** alpha image from this flyout.

9. Choose the **Color** palette. In this palette, set the values of the **R**, **G**, and **B** sliders to **160**, **85**, and **80**, respectively. Set the value of the **Rgb Intensity** slider in the top shelf to **20**. Next, drag the cursor to the white portion of the eyeball to create a pattern, as shown in Figure 8-15. Similarly, create more such patterns on the white area of the eyeball, refer to Figure 8-16.

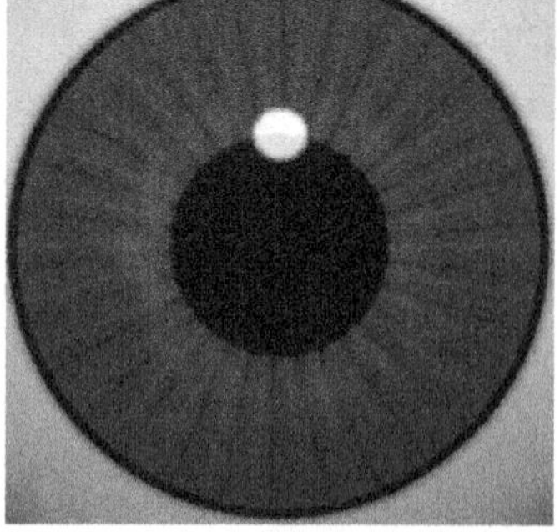

Figure 8-14 Detail added to the iris

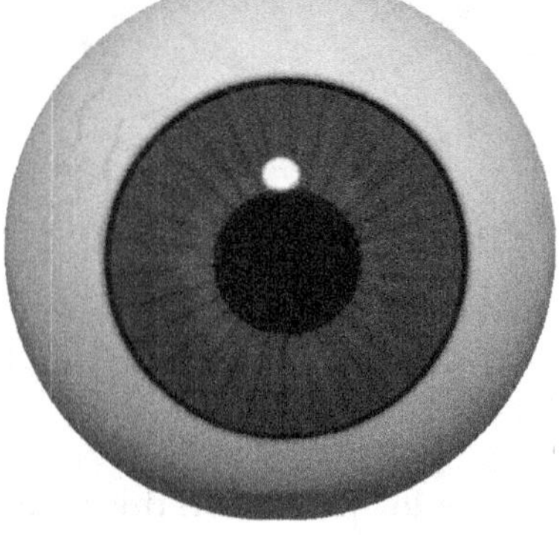

Figure 8-15 A pattern created on the white area of the eyeball

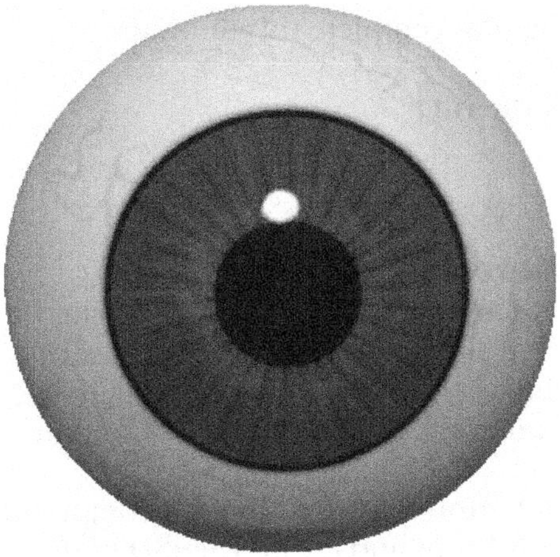

Figure 8-16 *More patterns created on the white area of the eyeball*

Saving the Model

In this section, you will save the file using the steps given next.

1. Choose the **Save As** button from the **Tool** palette; the **Save ZTool** dialog box is displayed. In this dialog box, browse to the location *\Documents\ZBrushprojects\c08*.

2. Enter **c08tut1** in the **File name** edit box and then choose the **Save** button.

Tutorial 2

In this tutorial, you will create the model of a ring using different materials. The final output of the model is shown in Figure 8-17. **(Expected time: 30 min)**

Figure 8-17 *The model of the ring*

The following steps are required to complete this tutorial:

a. Create the basic shape of the ring.
b. Add details to the ring.
c. Save the model.

Creating the Basic Shape of the Ring

In this section, you will create the basic shape of the ring using the **Ring3D** primitive.

1. Choose the **Init ZBrush** button from the **Preferences** palette; the message box is displayed. Choose the **Yes** button from the message box; ZBrush is initialized to its default state.

2. Choose the Current Tool button from the **Tool** palette; a flyout is displayed. Choose the **Ring3D** primitive from this flyout and create **Ring3D** in the canvas. Next, choose the **Edit** button from the top shelf to switch to the edit mode.

3. Expand the **Initialize** subpalette in the **Tool** palette. In this subpalette, set the value of the **SRadius** slider to **14**, refer to Figure 8-18; the shape of the ring changes, refer to Figure 8-19.

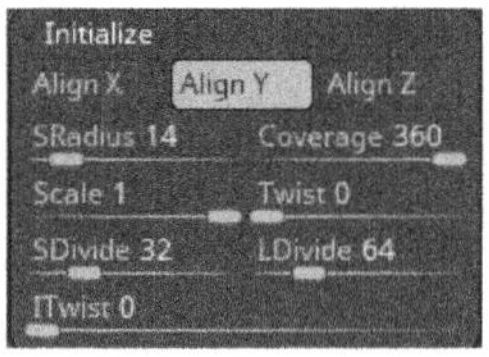

Figure 8-18** The value of the **SRadius** slider set to **14

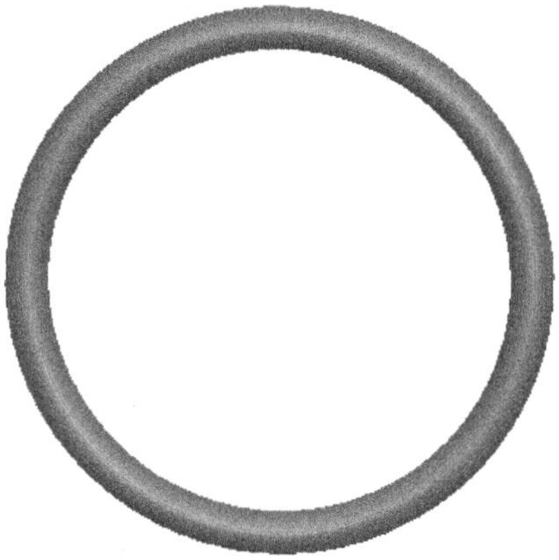

***Figure 8-19** The shape of the ring changed*

4. Choose the **Make PolyMesh3D** button from the **Tool** palette to convert the primitive into a polymesh.

5. Set the value of the **SDiv** slider in the **Geometry** subpalette to **5** by choosing the **Divide** button four times.

6. Choose the Current Brush button from the left shelf; a flyout containing different sculpting brushes is displayed. Choose the **MaskRect** brush from this flyout; a message box is displayed prompting you to press CTRL to use this brush. Choose the **OK** button to close this message box.

7. Rotate the canvas such that you can view the side view of the ring, refer to Figure 8-20.

8. Press and hold the CTRL key and drag the cursor on the surface of the ring to create a rectangular stroke, as shown in Figure 8-91; a mask is created on the surface of the ring.

9. Press CTRL+I; the mask is inverted. Next, expand the **Deformation** subpalette in the **Tool** palette.

   Next, you will decrease the size of the unmasked area along the Y-axis using the **Size** slider.

10. Deactivate the **x** and **z** options corresponding to the **Size** slider by choosing the **x** and **z** buttons located on the right side of the slider.

11. Set the value of the **Size** slider to **-5**; the unmasked area is pushed in. Next, press and hold the CTRL key and then drag the cursor in the canvas area; the mask is removed from the ring, refer to Figure 8-21.

   Now, you need to download a material from the MatCap library of the Pixologic website *https://pixologic.com/zbrush/downloadcenter/library/*. The process of downloading materials has already been discussed.

Figure 8-20 *A rectangular stroke created on the surface of the ring*

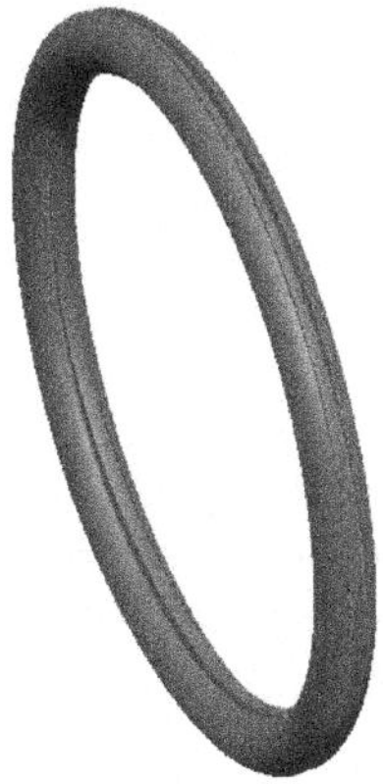

Figure 8-21 *The mask removed from the ring*

12. Download the **Gold 1** material from the **METAL** category of the **MatCap** library and extract the material from the zip file.

13. Choose the **Material** palette. In this palette, choose the **Load** button; the **Load Material** dialog box is displayed. In this dialog box, browse to the location where you have saved the material. Select the *sl_gold_1.ZMT* file and then choose the **Open** button in the **Load Material** dialog box; the gold material is applied to the ring, refer to Figure 8-22.

14. Choose the **Material** palette. In this palette, expand the **Modifiers** subpalette. In this subpalette, set the value of the **Intensity A** slider to **5**, refer to Figure 8-23; the ring becomes brighter.

Figure 8-22 *The gold material applied to the ring*

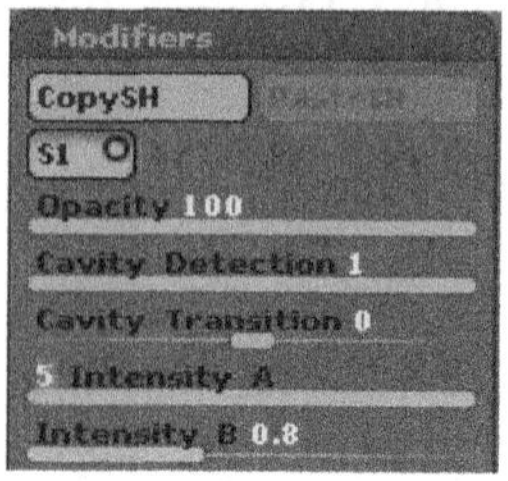

Figure 8-23 *The value of the **Intensity A** slider set to **5***

Adding Details to the Ring

In this section, you will add details to the ring using Insert brushes and different materials.

1. Choose the **Mrgb** button from the top shelf and make sure the **Zadd** button is chosen. Next, choose the **Color** palette. In this palette, choose the **FillObject** button.

2. Choose the Current Material button from the left shelf; a flyout containing different materials is displayed. In this flyout, choose the **ReflectedMap2** material from the **Standard Materials** area.

3. Choose the **Material** palette. In this palette, expand the **Modifiers** subpalette. In this subpalette, set the value of the **Ambient** and **Specular** sliders to **100** and the value of the **Diffuse** slider to **50**, refer to Figure 8-24.

4. Choose the **Transform** palette. In this palette, choose the **Activate Symmetry** button. Next, choose the **(R)** button; the **RadialCount** slider is activated. By default, the **>X<** button is chosen. To deactivate this button, choose it again. Next, choose the **>Z<** button; the radial symmetry in Z-axis is activated. Set the value of the **RadialCount** slider to **100**.

   Next, you will create diamonds in the ring using the **InsertHSphere** brush. Before using this brush, you have to make sure the geometry has not been subdivided.

5. Expand the **Geometry** subpalette in the **Tool** palette. In this subpalette, choose the **Del Lower** button, refer to Figure 8-25.

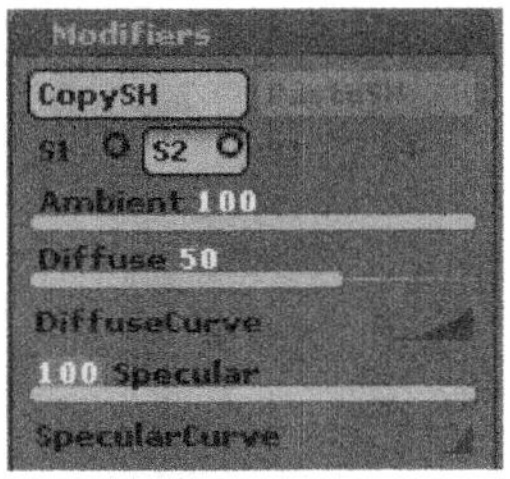

Figure 8-24 *The values of the* ***Ambient*** *and* ***Specular*** *sliders set to* ***100***

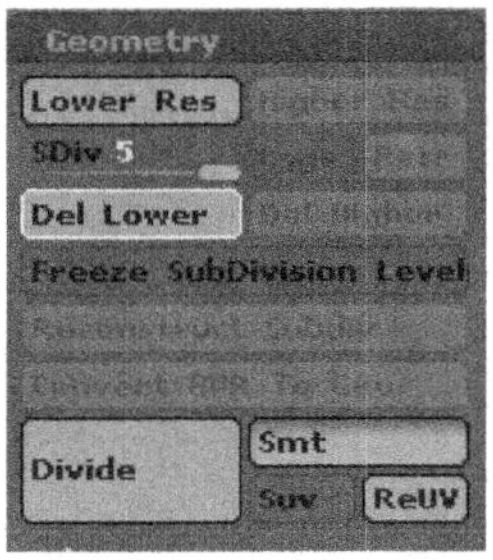

Figure 8-25 *The* ***Del Lower*** *button chosen in the* ***Geometry*** *subpalette*

6. Choose the **InsertHSphere** brush from the **Brush** tab of the **LightBox** browser. Next, drag the cursor on the recessed area of the ring; small diamonds are created in the recessed area of the ring, refer to Figure 8-26.

7. Choose the **Mrgb** button in the top shelf. Choose the **Color** palette again. In this palette, choose the **FillObject** button. Remove the mask from the ring by pressing and holding the CTRL key and then dragging the cursor in the canvas area.

8. Rotate the view of the canvas such that the ring is snapped horizontally to the canvas, refer to Figure 8-27.

Figure 8-26 *Diamonds created in the recessed area of the ring*

Figure 8-27 *Ring snapped horizontally with the canvas*

9. Choose the Current Material button from the left shelf; a flyout containing different materials is displayed. In this flyout, choose the **sl_gold_1** material.

10. Choose the **Transform** palette. In this palette, choose the **Activate Symmetry** button to deactivate symmetry.

11. Choose the Current Brush button from the left shelf; a flyout containing different sculpting brushes is displayed. Choose the **IMM SteamGears** brush from this flyout. Next, drag the cursor to the center of the ring; a golden mesh is inserted in the ring, refer to Figure 8-28.

12. Rotate the view of the canvas and then position the newly created mesh using the action line of the **Rotate** button, refer to Figure 8-29.

Figure 8-28 *A golden mesh inserted in the ring*

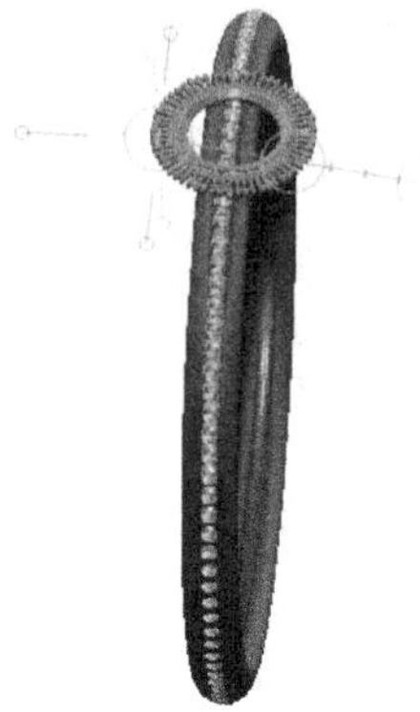

Figure 8-29 *The golden mesh adjusted using the action line*

13. Increase the size of the mesh along the Z-axis using the **Size** slider of the **Deformation** subpalette, refer to Figure 8-30.

14. Make sure the **Draw** button is chosen in the top shelf. Make sure the **Mrgb** and **Zadd** buttons are chosen in the top shelf. Next, choose the **Color** palette. In this palette, choose the **FillObject** button.

15. Choose the Current Material button from the left shelf; a flyout containing different materials is displayed. In this flyout, choose the **ReflectedMap2** material.

16. Choose the **Color** palette. In this palette, set the values of the **R**, **G**, and **B** sliders to **155**, **5**, and **5**, respectively.

17. Choose the **InsertHSphere** brush from **Insert-IMM** in the **Brush** tab of the **LightBox** browser; Next, drag the cursor inside the mesh created; a red colored stone is created.

18. Increase the size of the mesh along the Z-axis using the **Size** slider of the **Deformation** subpalette such that it fits inside the golden mesh, refer to Figure 8-31. Remove the mask from the ring by pressing and holding the CTRL key and then dragging the cursor in the canvas area.

Figure 8-30 *The size of the golden mesh increased along the Z-axis*

Figure 8-31 *The red stone fitted inside the golden mesh*

You can also create a ring box and place the ring inside it.

19. Expand the **SubTool** subpalette of the **Tool** palette. In this subpalette, choose the **Append** button; a flyout is displayed. In this flyout, choose the **Cube3D** primitive; a cube is created in the canvas.

20. Choose the thumbnail for the cube from the list of subtools. Set the value of **SDiv** slider in the **Geometry** subpalette to **5** by choosing the **Divide** button four times; the cube becomes smoother.

21. Modify the shape of the cube using the modeling techniques discussed in the previous chapters to create a ring box. Next, assign a red colored plastic material to it, refer to Figure 8-32.

Figure 8-32 *The ring placed inside the ring box*

Saving the Model

In this section, you will save the file using the steps given next.

1. Choose the **Save As** button from the **Tool** palette, the **Save ZTool** dialog box is displayed. In this dialog box, browse to the location *\Documents\ZBrushprojects\c08*.

2. Enter **c08tut2** in the **File name** edit box and then choose the **Save** button.

EXERCISES

The output of the models used in the following exercises can be accessed by downloading the *c08_ZBrush_2018_exr.zip* file from *www.cadsofttech.com*. The path of the file is as follows: *Textbooks > Animation and Visual Effects > ZBrush > Pixologic ZBrush 2018 for Novices*.

Exercise 1

Create the model of a cake using the radial symmetry and apply different materials to it, as shown in Figure 8-33. **(Expected time: 20 min)**

Figure 8-33 *Model of a cake*

Exercise 2

Create the model of a hut and apply different materials to it, as shown in Figure 8-34. **(Expected time: 30 min)**

Figure 8-34 *Model of a hut*

Chapter 9

Texturing in ZBrush

Learning Objectives

After completing this chapter, you will be able to:

- *Work with the Texture palette*
- *Work with the Spotlight tool*
- *Apply textures on different objects*

INTRODUCTION

In this chapter, you will learn to assign various textures to 3D meshes created in ZBrush. These textures give a realistic look to the objects. ZBrush consists of a library of textures that can either be accessed from the Current Texture button located in the left shelf or the **LightBox** browser.

TEXTURES IN ZBrush

ZBrush consists of an extensive library of textures. To access these textures, choose the Current Texture button from the left shelf; a flyout containing different types of textures will be displayed, refer to Figure 9-1. This flyout consists of two areas namely **Quick Pick** and **Textures**. The **Quick Pick** area consists of the recently chosen textures and the **Textures** area consists of all the textures available in ZBrush. Besides this, there are some additional textures in ZBrush that can be accessed from the **Texture** tab of the **LightBox** browser, refer to Figure 9-2.

Figure 9-1 *The flyout displayed on choosing the Current Texture button*

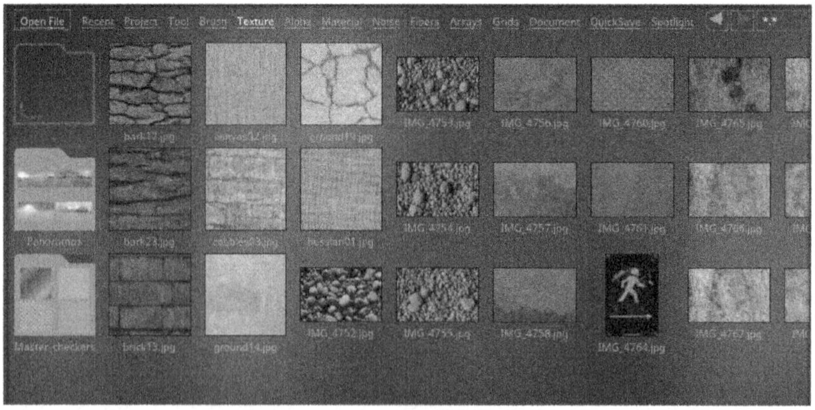

Figure 9-2 *The textures displayed in the* ***Texture*** *tab of the LightBox browser*

The flyout displayed on choosing the Current Texture button consists of different buttons located at its bottom. These buttons are discussed next.

Import

The **Import** button is used to import a new texture file to ZBrush. On choosing this button, the **Import Image** dialog box will be displayed. You can download additional textures from the Pixologic site by visiting the following URL: *https://pixologic.com/zbrush/downloadcenter/texture/*. This site has different categories of textures displayed under the **Texture Library** tab, refer to Figure 9-3. To download a texture from this site, choose the desired category under the **Texture Library** tab. For instance, to download different bricks textures, choose the **BRICK** category from the library; different textures of bricks will be displayed, refer to Figure 9-4.

***Figure 9-3** The textures in the **Texture Library** tab*

***Figure 9-4** Different brick texture displayed*

Select the desired texture; a preview of that texture will be displayed, refer to Figure 9-5. Next, choose the **DOWNLOAD HERE** button located on the right side of the preview, refer to Figure 9-5. The material will be downloaded as a zip file. Next, you need to extract the material from the zip file.

Figure 9-5 *Preview of the brick texture*

After extracting the textures, you can import them to the ZBrush interface. To do so, choose the **Import** button. On choosing this button, the **Import Image** dialog box will be displayed. In this dialog box, browse to the location where the downloaded texture images are saved. Choose the desired texture image and then choose the **Open** button; the new texture will be displayed in the flyout.

Export

The **Export** button is used to export a texture image to another location on your computer. By default, this button is not activated. To activate this button, you need to choose a texture image from the flyout displayed on choosing the Current Texture button.

You can export the texture image as a *.psd* file. You can modify the exported texture in Photoshop and then import it again to ZBrush.

Clone

The **Clone** button is used to create a copy of a chosen texture image.

MakeAlpha

The **MakeAlpha** button is used to convert a texture into a gray scale alpha image. To do so, choose the Current Texture button; a flyout containing different texture images will be displayed. In this flyout, choose the **Texture 01** texture image. Next, choose the **MakeAlpha** button; the texture will be displayed as a gray scale alpha image in the Current Alpha button.

FillLayer

The **FillLayer** button is used to fill the canvas in a particular layer with a color, material or a texture. On choosing this button, the objects in a layer will be deleted. To fill the canvas in a particular layer with a color, choose the Current Texture button, a flyout containing different textures will be displayed. From this flyout, choose the **Texture 01** texture image. Now, choose the Current Material button, a flyout containing different materials will be displayed. In this flyout, choose the **ReflectRed** material from the **MatCap Materials** area; the object in the canvas will be filled with the material. Choose the Current Texture button again, a flyout will be displayed. In this flyout, choose the **FillLayer** button, refer to Figure 9-6.

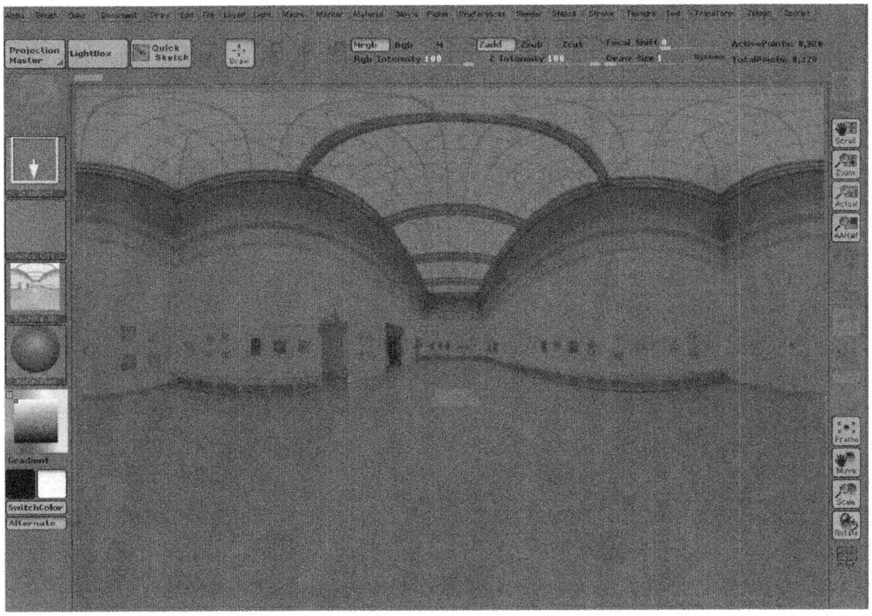

Figure 9-6 *The canvas filled with the texture and the material*

Cd

The term Cd stands for Clear Depth. The **Cd** button is used to determine whether the canvas will be filled with the selected texture or not. If this button is not chosen, then the canvas will not be filled with a texture on choosing the **FillLayer** button.

CropAndFill

The **CropAndFill** button is used to crop a document according to the dimensions of the selected texture and fills it with that texture. For example, to crop a document and fill it with **Texture 10**, choose the Current Texture button; a flyout will be displayed. In this flyout, choose the **Texture 10** texture image. Next, choose the **CropAndFill** button, refer to Figure 9-7.

Before choosing this button, make sure the **Cd** button is chosen. If the **Cd** button is not chosen before choosing the **CropAndFill** button, the document will be cropped according to the dimensions of the selected texture, but it will not be filled with the texture.

GrabDoc

The **GrabDoc** button is used to capture the canvas and add this capture as a texture in the flyout displayed on choosing the Current Texture button. To add the capture as a texture, choose the Current Texture button; a flyout will be displayed. In this flyout, choose the **GrabDoc** button; a new texture image with a name **ZGrab01** will be displayed in the flyout, refer to Figure 9-8.

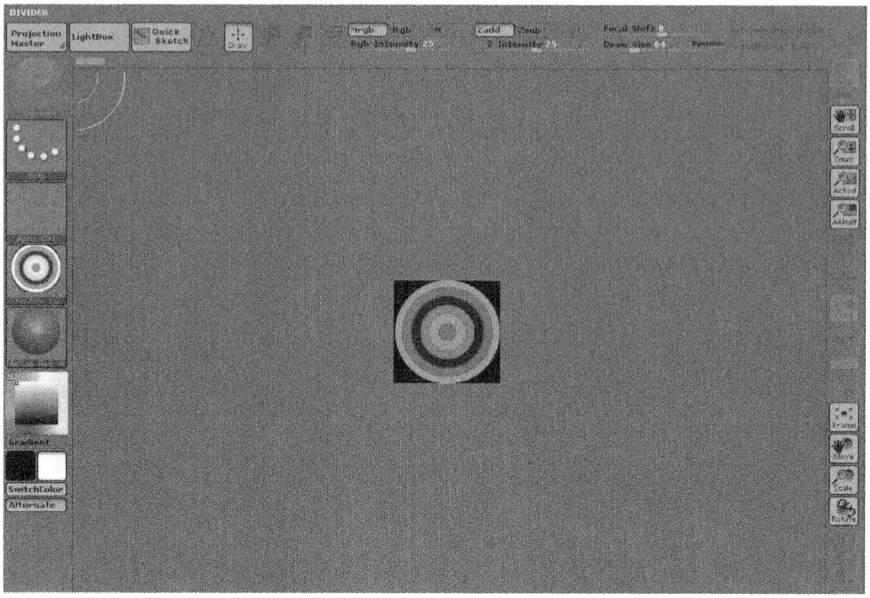

*Figure 9-7 The document cropped and filled with **Texture 10***

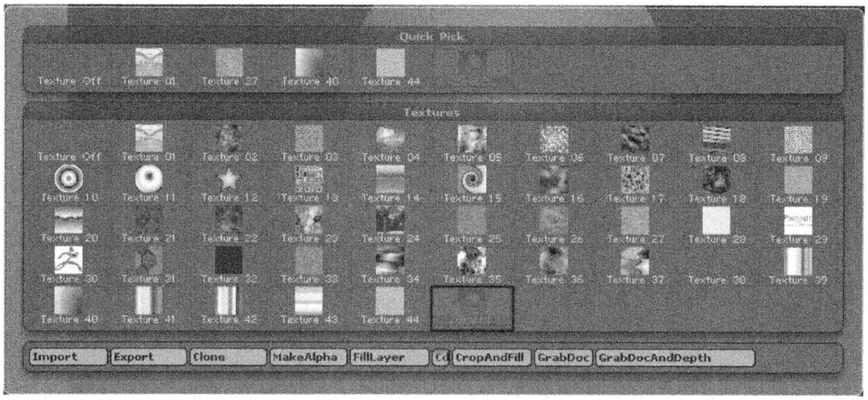

*Figure 9-8 A new texture image with the name **ZGrab01** displayed in the flyout*

GrabDocAndDepth

The **GrabDocAndDepth** button is used to capture the canvas and add this capture as a texture in the flyout displayed on choosing the Current Texture button. These textures can then be exported in the *.png* format containing transparency or can also be exported as alpha images containing depth.

Texture Palette

The **Texture** palette consists of different options that enable you to open and modify texture images. The **Texture** palette is located at the top of the interface. On choosing this palette, it will be expanded and different buttons and sliders will be displayed, refer to Figure 9-9.

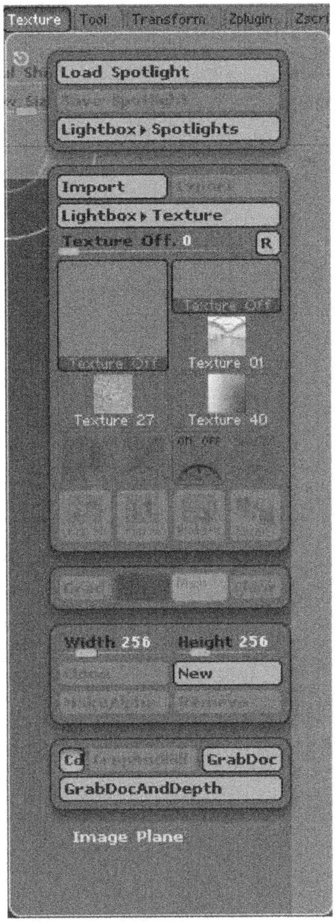

Figure 9-9 The ***Texture*** *palette*

ZProject

The **ZProject** brush is mainly used for texturing. By using the **ZProject** brush, you can transfer the texture from the reference image into a 3D model. The **ZProject** brush uses the Z axis of the canvas to transfer the texturing details from the reference image into a 3D model. To transfer the texture from the reference image into a 3D model, make sure a sphere in the canvas and then choose the **Edit** button from the top shelf. Next, convert the sphere into a polymesh by choosing the **Make PolyMesh3D** button in the **Tool** palette.

To view the texture on the sphere clearly, you need to subdivide the geometry. To do so, expand the **Geometry** subpalette in the **Tool** palette. From this subpalette, set the value of the **SDiv** slider to **5** by clicking on the **Divide** button four times. Next, choose the Current Material button; a flyout will be displayed. From this flyout, choose the **SkinShade4** material; the material will be applied to the sphere, refer to Figure 9-10. Expand the **Texture** palette. In this palette, choose the Current Texture button; a flyout will be displayed. In this flyout, choose the **Texture 42** texture image; the texture image will be displayed in the Current Texture button, refer to Figure 9-11.

Figure 9-10 *The material applied to the sphere*

Figure 9-11 *The texture image displayed in the Current Texture button*

Next, expand the **Image Plane** subpalette. In this subpalette, choose the **Load Image** button; the texture image will be displayed behind the sphere, refer to Figure 9-12. Choose the Current Brush button; a flyout containing different types of brushes will be displayed. In this flyout, choose the **ZProject** brush. Choose the **Zsub** button in the top shelf to deactivate it and make sure the **Rgb** button is chosen. Next, press and hold the left mouse button and drag the cursor on the sphere; the texture will be projected from the background texture image to the sphere, refer to Figure 9-13.

Before loading the texture image, you can also specify the image size as required by using the **Image Size** slider located in the **Image Plane** subpalette. To remove the image from the canvas, press CTRL+N.

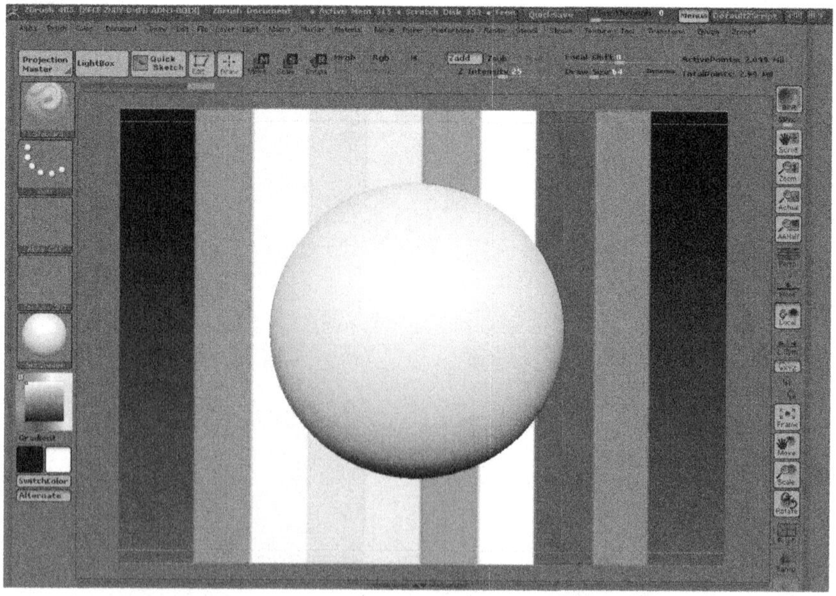

Figure 9-12 *The texture image displayed behind the sphere*

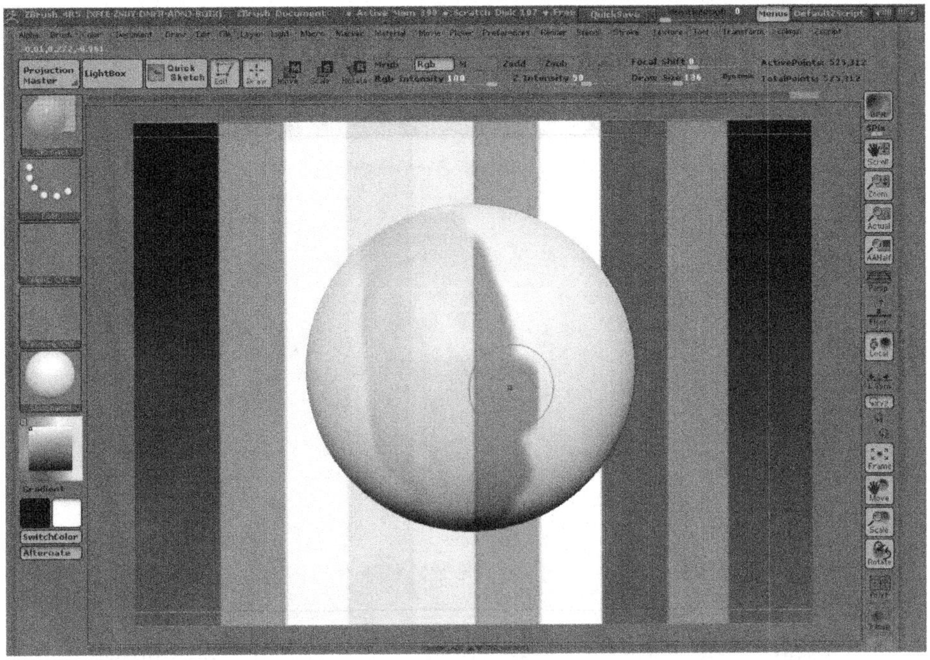

Figure 9-13 The texture projected from the texture image onto the sphere

Spotlight

The Spotlight tool is used to modify an existing texture image or project the details from an image into a 3D model. This tool consists of different options that enable you to manipulate an existing texture image as per your requirement and then apply that texture to a 3D model. The Spotlight tool can be invoked from the **Texture** palette. To load the Spotlight tool, you need to select a texture image first. You can either import a new texture image or use the existing default textures in ZBrush. To loading the Spotlight, expand the **Texture** palette. In this palette, choose the Current Texture button; a flyout will be displayed. In this flyout, choose the **Texture 01** texture image; the texture image will be displayed in the Current Texture button. You can also import a new texture image using the **Import** button.

*Figure 9-14 The **Add To Spotlight** button chosen*

Choose the **Add To Spotlight** button in the **Texture** palette, refer to Figure 9-14; the texture image will be displayed in the canvas. Besides the texture image, a ring containing different icons will be displayed on the texture image, refer to Figure 9-15.

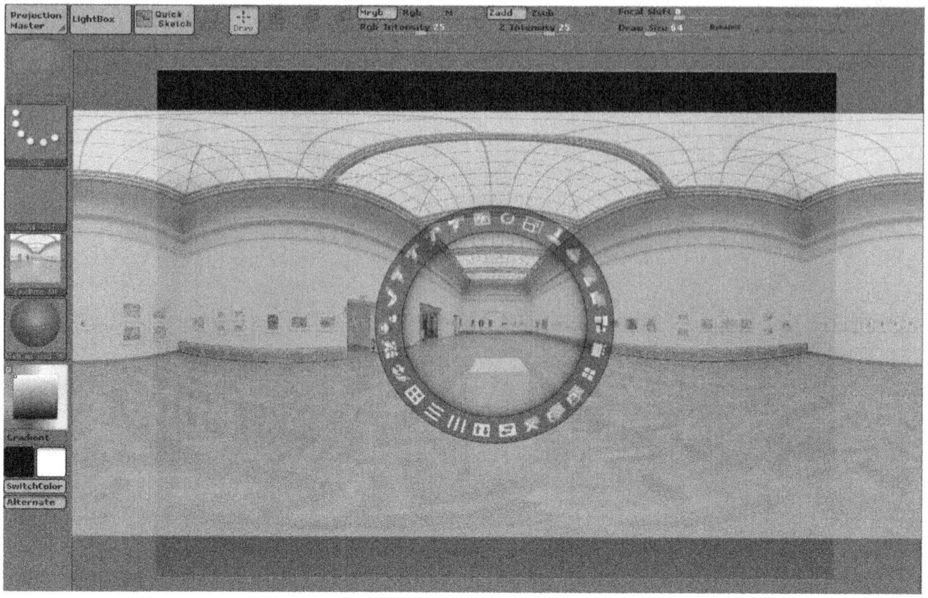

Figure 9-15 *The texture image and the ring displayed in the canvas*

POLYPAINTING

The polypainting technique allows you to paint colors directly on to the 3D objects using different sculpting brushes. To paint colors directly on to the 3D object, apply the **SkinShade4** material to 3D object.

In the **Tool** palette, expand the **Polypaint** subpalette. In this subpalette, choose the **Colorize** button, refer to Figure 9-16. Next, choose the desired colors from the Current Color swatch and using the different brushes and alphas, start painting on the surface of the sphere, refer to Figure 9-17. If you want to add depth along with the color, make sure the **Zadd** button is chosen in the top shelf.

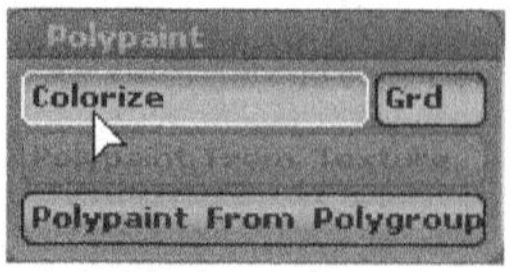

Figure 9-16 *The* ***Colorize*** *button chosen*

Figure 9-17 *Different colors painted on the surface of the sphere*

TUTORIALS

All the files used in the tutorial can be downloaded from the CADSoft website (*www.cadsofttech.com*). These files are compressed in zip file format and are required to be extracted before using them in the tutorials. The path of the files is as follows: *Textbooks > Animation and Visual Effects > Pixologic ZBrush 2018 > Pixologic ZBrush 2018 for Novices*

Tutorial 1

In this tutorial, you will apply different colors and textures on the surface of the hut. The final output of the model is shown in Figure 9-18. **(Expected time: 20 min)**

Figure 9-18 *The textured model of the hut*

The following steps are required to complete this tutorial:

a. Apply wooden texture on the walls of the hut.
b. Apply textures on the door, windows, and roof of the hut.
c. Apply textures on the remaining parts of the hut.
d. Save the model.

Applying Wooden Texture on the Walls of the Hut

In this section, you will apply a wooden texture on the walls of the hut using Spotlight.

1. Choose the **Init ZBrush** button from the **Preferences** palette; the message box is displayed. Choose the **Yes** button from the message box; ZBrush is initialized to its default state.

2. Choose the **Tool** palette. In this palette, choose the **Load Tool** button; the **Load ZTool** dialog box is displayed. In this dialog box, browse to *\Documents\ZBrushprojects\c09* and select the **c09tut1start.ZTL** file in it. Next, choose the **Open** button in this dialog box.

3. Press and hold the left mouse button and drag the cursor in the canvas area; the scene is displayed in the canvas. Next, choose the **Edit** button from the top shelf to switch to the edit mode. Rotate the view of the canvas, refer to Figure 9-19.

Figure 9-19 *The untextured model of the hut*

4. Choose the Current Material button from the left shelf; a flyout containing different materials is displayed. In this flyout, choose the **RGB Levels** material from the **Standard Materials**; the material is applied to the hut.

5. Choose the Current Brush button from the left shelf; a flyout containing different sculpting brushes is displayed. Choose the **MaskRect** brush from this flyout; a message box prompting you to press CTRL to use this brush is displayed. Choose the **OK** button to close this message box.

6. Press and hold the CTRL key and drag the cursor on the door and window of the hut to create a rectangular mask, refer to Figure 9-20.

7. Choose the Current Brush button from the left shelf; a flyout containing different sculpting brushes is displayed. Choose the **MaskCircle** brush from this flyout; a dialog box prompting you to press CTRL to use this brush is displayed. Choose the **OK** button to close this dialog box.

8. Press and hold the CTRL key and choose the Current Stroke button; a flyout is displayed. In this flyout, choose the **Square** button located at the bottom of the flyout. Similarly, choose the **Center** button. Next, release the CTRL key.

9. Press and hold the CTRL key and drag the cursor on the surface of hut to create two circular masks on the ventilators of the hut, refer to Figure 9-20.

   As the symmetry in X-axis is activated in this model by default; the mask will be created on the right side also. Remove the mask on the right hand side by pressing CTRL+ALT and then drag the mouse.

10. Choose the Current Brush button from the left shelf; a flyout containing different sculpting brushes is displayed. Choose the **MaskCurve** brush from this flyout; a message box prompting you to press CTRL to use this brush is displayed. Choose the **OK** button to close this dialog box.

11. Press and hold the CTRL key and drag the cursor in an upward direction to create a stroke, as shown in Figure 9-21; a mask is applied on the roof of the hut. Similarly, create a mask on the top portion of the hut, refer to Figure 9-22.

Figure 9-20 *Masks created on the door, windows, and ventilators of the hut*

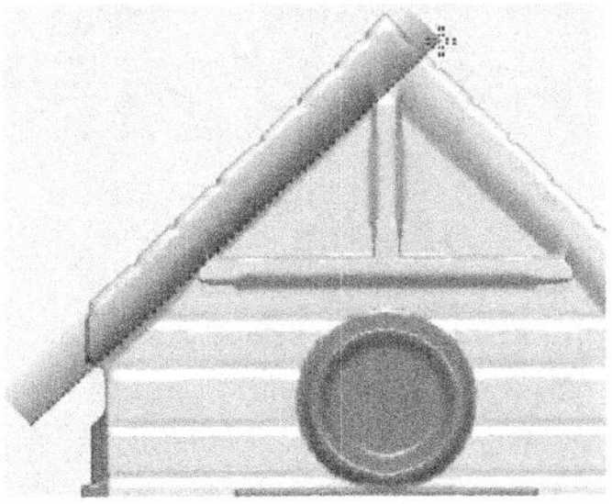

Figure 9-21 *Mask created on the roof of the hut using the **MaskCurve** brush*

12. Expand the **Texture** palette. In this palette, choose the **LightBox Texture** button, refer to Figure 9-23; different textures are displayed in the **LightBox** browser.

13. Hover the cursor on the LightBox browser and drag it toward left to view all the textures available in the browser. Choose the **wood08.jpg** texture image, refer to Figure 9-24, and double-click on it; the image is displayed in the Current Texture button in the **Texture** palette.

14. Choose the **Add To Spotlight** button in the **Texture** palette; the image is added to the Spotlight and is displayed in the canvas.

Figure 9-22 Mask created on the top area of the hut

Figure 9-23 *The* ***LightBox Texture*** *button chosen*

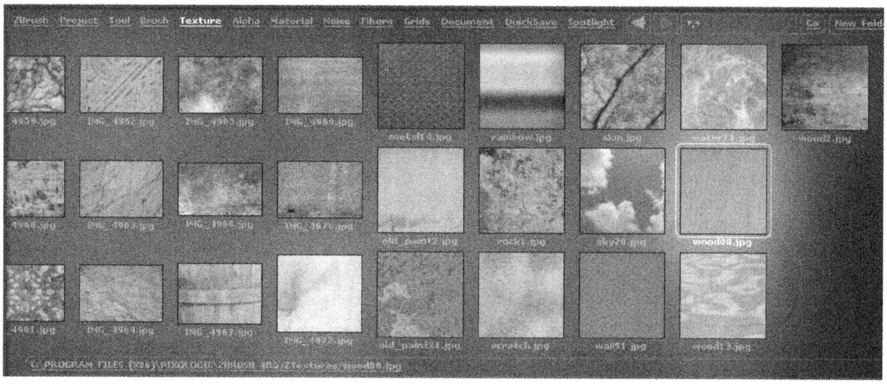

Figure 9-24 *The* ***wood08.jpg*** *texture image chosen from the LightBox browser*

15. Move and scale the image such that it covers the hut. In the Spotlight ring, hover the cursor at the **Rotate** icon and drag it in the clockwise direction; the image is rotated, refer to Figure 9-25.

    The image should be rotated in such a way that the pattern on the image is displayed horizontally.

16. Press Z; the ring disappears from the canvas. Next, choose the **Rgb** button from the top shelf and make sure the **Zadd** button is not chosen. Drag the cursor on the texture on the image; the texture is projected on the surface of the hut.

17. Rotate the view of the canvas such that you can view the side and back views of the hut and project the texture on all the views of the hut. To view the texture on the hut, press SHIFT+Z; the texture image disappears and the texture is projected on the unmasked areas of the hut, refer to Figure 9-26.

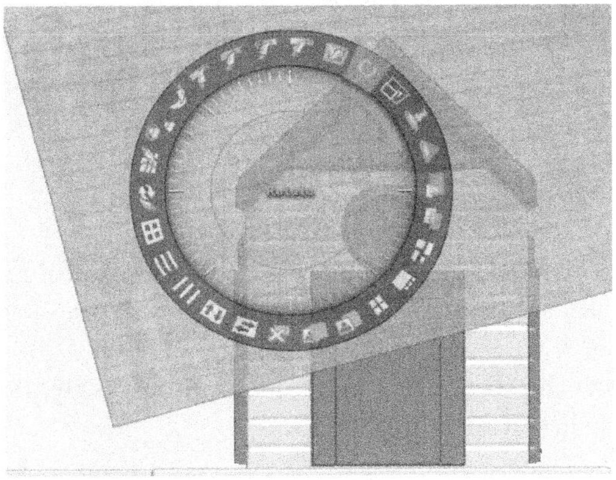

Figure 9-25 The texture image rotated in the canvas

Figure 9-26 The texture projected on the unmasked areas of the hut

18. Press CTRL+I; the mask is inverted. Next, create a mask on the door using the **MaskRect** brush, refer to Figure 9-27.

19. Press Z again; the image is displayed in the canvas. Rotate the image, as shown in Figure 9-69, using the **Rotate** icon of the Spotlight ring. Press Z; the ring disappears. Next, drag the cursor on the door frame; the texture is projected on the door frame. Next, press SHIFT+Z to view the texture on the hut.

20. Remove the mask from the hut by pressing and holding the CTRL key and dragging the cursor in the canvas area, refer to Figure 9-28.

Figure 9-27 The texture image rotated in the canvas

Figure 9-28 The mask removed from the surface of the hut

Applying Textures on the Door, Windows, and Roof of the Hut

In this section, you will apply textures on the door, windows, and roof of the hut using the Spotlight.

1. Create a mask on the door of the hut using the **MaskRect** brush. Next, press CTRL+I; the mask is inverted.

2. Choose the **IMG_4850.jpg** texture image from the **LightBox** browser and then double-click on it; the image is displayed in the Current Texture button of the **Texture** palette.

3. Choose the **Add To Spotlight** button from the **Texture** palette; the image is displayed in the canvas. Scale and move the image, as shown in Figure 9-29.

4. Press Z; the ring disappears. Next, drag the cursor on the door of the hut; the texture is projected on its surface. Press SHIFT+Z to hide the texture image and then remove the mask from the surface of the hut, refer to Figure 9-30.

Figure 9-29 *The texture image moved and rotated in the canvas*

Figure 9-30 *The mask removed from the surface of the hut*

5. Expand the **Color** palette. In this palette, set the values of the **R**, **G**, and **B** sliders to **100**, **32**, and **15**, respectively, as shown in Figure 9-31.

6. Choose the Current Brush button from the left shelf; a flyout containing different brushes is displayed. In this flyout, choose the **MaskCircle** brush; a message box prompting you to press CTRL to use this brush is displayed. Choose the **OK** button to close this message box.

7. Press and hold the CTRL key and choose the Current Stroke button; a flyout is displayed. In this flyout, choose the **Square** button located at the bottom of the flyout. Similarly, choose the **Center** button. Next, release the CTRL key.

8. Press and hold the CTRL key and drag the cursor on the circular ventilators of the hut. Remove the mask from the inner part of the ventilator by pressing CTRL+ALT, refer to Figure 9-32. Make sure the mask is removed from the backside of the hut.

9. Press CTRL+I; the mask is inverted. Make sure the **Rgb** button is chosen in the top shelf. Drag the cursor on the surface of the frame of the ventilator to apply paint on it, refer to Figure 9-33. Remove the mask from the hut by pressing and holding the CTRL key and dragging the cursor in the canvas area.

10. Apply mask on the entire hut and then remove masks from the window frame and the ventilator frame on the left side of the hut. Next, drag the cursor on the unmasked areas of the hut to paint them, refer to Figure 9-34.

Figure 9-31 *The values of the* ***R****,* ***G****, and* ***B*** *sliders set*

Figure 9-32 *Mask removed from the inner part of the ventilator*

Figure 9-33 *The frame of the ventilator painted*

Figure 9-34 *The frame of the window painted*

11. Press and hold the CTRL key and then drag the cursor in the canvas area; mask is removed from the hut, refer to Figure 9-35.

12. Create a mask on the window panes and the inner area of the ventilator, refer to Figure 9-36. Next, press CTRL+I to invert the mask.

Figure 9-35 *Mask removed from the hut*

Figure 9-36 *The mask applied on the window panes and the inner area of the ventilators*

13. Choose the **IMG_4801.jpg** texture image from the **LightBox** browser and then double-click on it; the image is displayed in the Current Texture button of the **Texture** palette.

14. Choose the **Add To Spotlight** button from the **Texture** palette; the image is displayed in the canvas, refer to Figure 9-37.

15. Scale up the texture image such that the glass window area in the texture image covers the window area of the hut.

16. Press Z; the ring disappears. Next, drag the cursor on the window panes of the hut; the texture is projected on its surface. Press SHIFT+Z to hide the texture image; the texture is displayed on the window panes, refer to Figure 9-38.

Figure 9-37 *The texture image displayed in the canvas*

Figure 9-38 *The texture projected on the window panes of the hut*

17. Choose the **IMG_4790.jpg** texture image from the **LightBox** browser and then double-click on it; the image is displayed in the Current Texture button of the **Texture** palette.

18. Choose the **Add To Spotlight** button from the **Texture** palette; the image is displayed in the canvas, refer to Figure 9-39.

19. Scale down the texture image such that it covers the ventilator of the hut.

20. Press Z; the ring disappears. Next, drag the cursor on the inner area of the ventilators; texture is projected on their surface. Press SHIFT+Z to hide the texture image; the texture is displayed on the ventilators, refer to Figure 9-40.

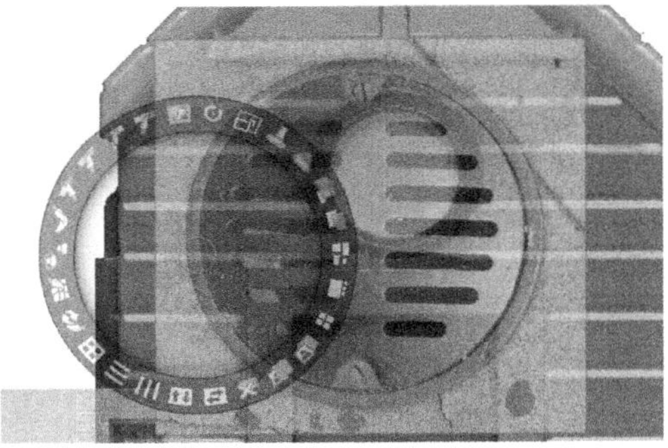

Figure 9-39 *The texture image displayed in the canvas*

Figure 9-40 *The texture displayed on the ventilators*

21. Create a mask on the roof of the hut using the **MaskCurve** brush and then create a mask on the window frame at the top of the hut using the **MaskRect** brush. Next, press CTRL+I; the mask gets inverted, refer to Figure 9-41.

22. Make sure the **Rgb** button is chosen in the top shelf. Drag the cursor on the top area of the hut to apply paint on it, as shown in Figure 9-42.

Figure 9-41 *The inverted mask*

Figure 9-42 *Paint applied on the top area of the hut*

23. Press CTRL+I to invert the mask. Expand the **Color** palette. In this palette, set the values of the **R**, **G**, and **B** sliders to **14**, **9**, and **9**, respectively. Drag the cursor on the top window of hut to apply paint on it, as shown in Figure 9-43.

Applying Textures on the Remaining Parts of the Hut

In this section, you will apply textures on the remaining parts of the hut.

1. Press SHIFT+Z; the Spotlight ring is displayed in the canvas. Next, select the **wood08.jpg** texture image from the canvas and move it such that it covers the staircase area of the hut, refer to Figure 9-44.

Figure 9-43 *The color painted on the top windows*

Figure 9-44 *The texture image placed on the staircase area*

2. Press Z; the ring disappears. Next, drag the cursor on the staircase area of the hut; the texture is projected on the staircase. Press SHIFT+Z to hide the texture image; the texture is displayed on the staircase, refer to Figure 9-45.

3. Choose the **brick13.jpg** texture image from the **LightBox** browser and then double-click on it; the image is displayed in the Current Texture button of the **Texture** palette.

4. Choose the **Add To Spotlight** button from the **Texture** palette; the image is displayed in the canvas. Move the image to the base of the hut and scale it down, as shown in Figure 9-46.

5. Choose the **Pin Spotlight** icon in the Spotlight ring; a bar is displayed on the icon, refer to Figure 9-46.

Figure 9-45 *The texture projected on the staircase area*

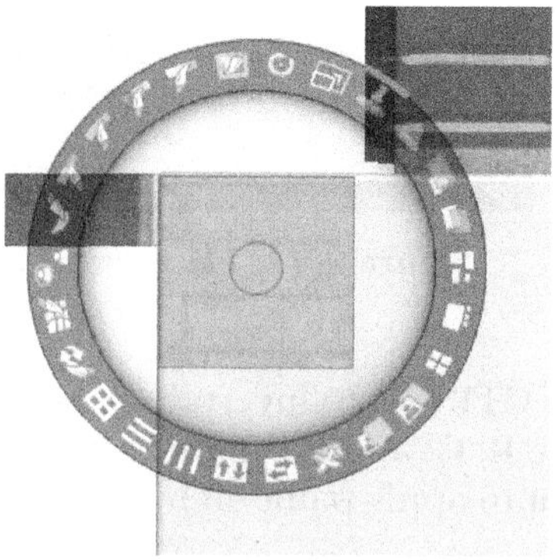

Figure 9-46 *The texture image moved and scaled down*

6. Press Z and drag the cursor on the texture image. Next, drag the cursor on the surface of the base; the texture is cloned on the front side of the base, refer to Figure 9-47.

7. Rotate the view of the canvas and continue dragging the cursor on all the sides of the base to project the brick texture on it, refer to Figure 9-48. Now, press SHIFT+Z to hide the texture image.

Figure 9-47 *The texture projected on the front side of the base*

Figure 9-48 *The texture projected on all the sides of the base*

8. Choose the **IMG_4819.jpg** texture image from the **LightBox** browser and then double-click on it; the image is displayed in the Current Texture button of the **Texture** palette.

9. Choose the **Add To Spotlight** button from the **Texture** palette; the image is displayed in the canvas.

10. Rotate the view of the canvas such that the top view of the hut is visible. Move the texture image such that it covers the floor of the base. Create a mask on the surface of the hut.

    Press Z and project the texture on the surface of the floor. Next, press SHIFT+Z to hide the texture image. Remove the mask from the hut; the texture is displayed on the floor, refer to Figure 9-49.

11. Create a mask on the top of the ground, refer to Figure 9-50.

Figure 9-49 *The texture projected on the floor of the base*

Figure 9-50 *A mask created on the top of the ground*

12. Expand the **FiberMesh** subpalette in the **Tool** palette. In this subpalette, choose the **LightBox Fibers** button; different types of fibers are displayed in the **LightBox** browser. Choose the **Fibers78.ZFP** file from the browser and double-click on it; the fibers are displayed on the ground.

13. Expand the **Modifiers** area in the **FiberMesh** subpalette. In this area, set the value of the **MaxFibers** slider to **209.4099** and then set the value of the **Length** slider to **93.042**, refer to Figure 9-51. Next, choose the **Accept** button; grass is displayed on the ground, refer to Figure 9-52.

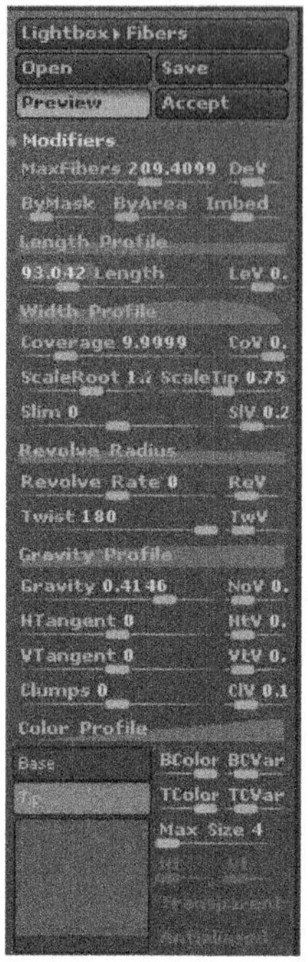

Figure 9-51 *The values of the sliders set*

Figure 9-52 *The grass displayed on the ground*

Saving the model

In this section, you will save the file using the steps given next.

1. Choose the **Save As** button from the **Tool** palette, the **Save ZTool** dialog box is displayed. In this dialog box, browse to the location *\Documents\ZBrushprojects\c09*.

2. Enter **c09tut1** in the **File name** edit box and then choose the **Save** button.

EXERCISE

The output of the model used in this exercise can be accessed by downloading the *c09_ZBrush_2018_exr.zip* file from *www.cadsofttech.com*. The path of the file is as follows: *Textbooks > Animation and Visual Effects > ZBrush > Pixologic ZBrush 2018 for Novices.*

Exercise 1

Download the file *c09_ZBrush_2018_exr.zip* from *www.cadsofttech.com*. Extract the contents of the zipped file and open *c09_exr1_start.ztl* from it, refer to Figure 9-53. Next, assign different textures to the scene using the LightBox textures, refer to Figure 9-54.

(Expected time: 20 min)

Figure 9-53 *Model of the hut without textures*

Figure 9-54 *Textured model of the hut*

Page intentionally left blank

Chapter 10

UV Master

Learning Objectives

After completing this chapter, you will be able to:

- *Work with the UV Master subpalette*
- *Unwrap different 3D models*

INTRODUCTION

In this chapter, you will work with the UV Master plugin. The UV Master plugin is used to unwrap the UVs in a 3D object by creating seams on its surface. Unwrapping a 3D object enables you to apply textures on the surface of that object accurately.

TUTORIALS

All the files used in the tutorial can be downloaded from the CADSoft website (*www.cadsofttech.com*). These files are compressed in zip file format and are required to be extracted before using them in the tutorials. The path of the files is as follows: *Textbooks > Animation and Visual Effects > Pixologic ZBrush 2018 for Novices*

Tutorial 1

In this tutorial, you will unwrap the model of the *DemoHead.ZTL* file present in the LightBox browser, refer to Figure 10-1. **(Expected time: 20 min)**

Figure 10-1 *The unwrapped model of the head*

The following steps are required to complete this tutorial:

a. Unwrap the model.
b. Adjust the seams in the model by control painting.
c. Save the unwrapped model.

Unwrapping the Model

In this section, you will unwrap the model using the default position of the seams.

1. Choose the **Init ZBrush** button from the **Preferences** palette; the message box is displayed. Choose the **Yes** button from the message box; ZBrush is initialized to its default state.

2. Choose the **DemoHead.ZTL** file from the **Tool** tab of the LightBox browser by double-clicking on it. Next, drag the cursor in the canvas area; the *DemoHead.ZTL* is displayed in the canvas.

3. Choose the **Edit** button in the top shelf to switch to the edit mode.

4. Expand the **Texture Map** subpalette in the **Tool** palette. In this subpalette, choose the **Texture Map** button; a flyout containing different texture images is displayed. In this flyout, choose the **Texture 03** texture image; a message box prompting you to apply UV mapping to the model is displayed, refer to Figure 10-2. Choose the **OK** button to close this message box.

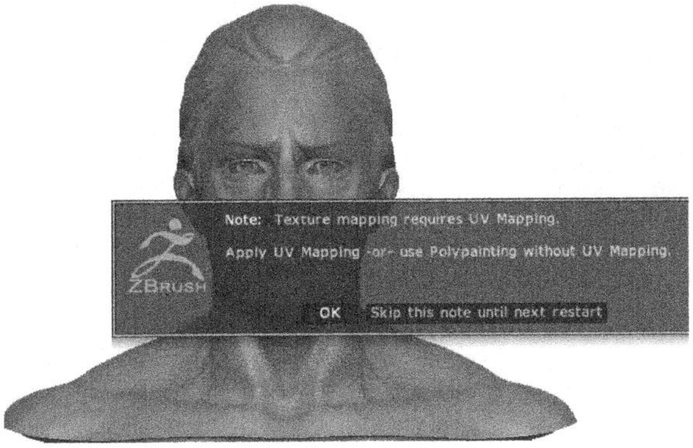

Figure 10-2 *The message box displayed on choosing the texture image*

5. Dock the **ZPlugin** palette in the right tray. In this palette, expand the **UV Master** subpalette. In this subpalette, choose the **Work On Clone** button; a clone of the model at its lowest subdivision level is displayed in the canvas, refer to Figure 10-3.

6. Choose the **Unwrap** button in the **UV Master** subpalette; the model is unwrapped and the **Texture 03** texture image is applied to it, refer to Figure 10-4. In this figure, you will notice that the checker pattern is distorted at the middle of the face.

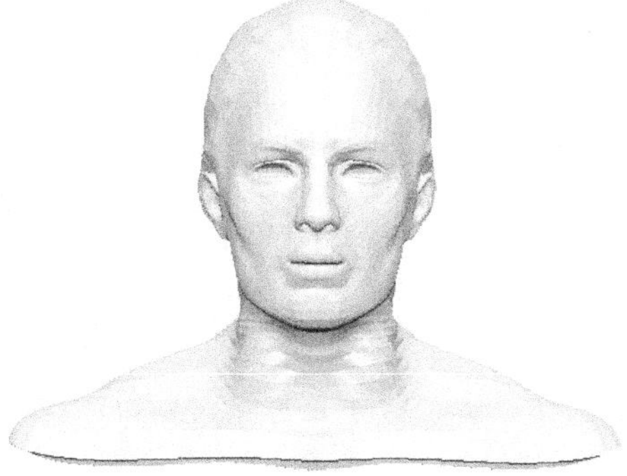

Figure 10-3 *The cloned model displayed in the canvas*

Figure 10-4 *The texture image applied to the model*

Adjusting the Seams in the Model by Control Painting

In this section, you will manually adjust the seams in the model using the **Protect** and **Attract** buttons.

1. In the **Texture Map** subpalette, choose the **Texture On** button to deactivate it, refer to Figure 10-5; the texture is removed from the model.

2. In the **UV Master** subpalette, choose the **CheckSeams** button; the seams are displayed on the model, refer to Figure 10-6.

Figure 10-5 *The* ***Texture On*** *button chosen*

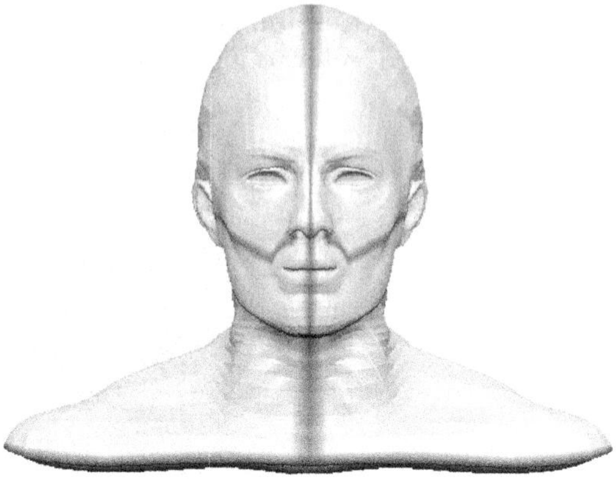

Figure 10-6 *The seams displayed on the model*

3. In the **UV Master** subpalette, choose the **Flatten** button; the flattened 2D image of the model is displayed, refer to Figure 10-7. In this figure, you will notice that the unwrapped image is not properly displayed.

4. Choose the **UnFlatten** button to display the 3D model again.

5. Choose the **Enable Control Painting** button in the **UV Master** subpalette. Next, choose the **Protect** button and then start painting on the front side of the face to remove seams from it, refer to Figure 10-8.

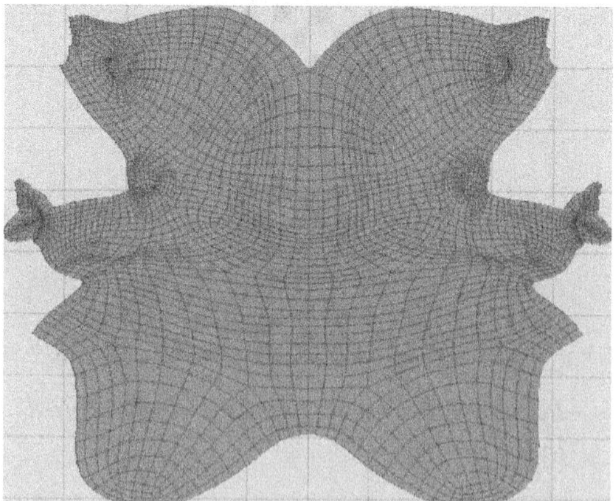

Figure 10-7 *The flattened image of the model*

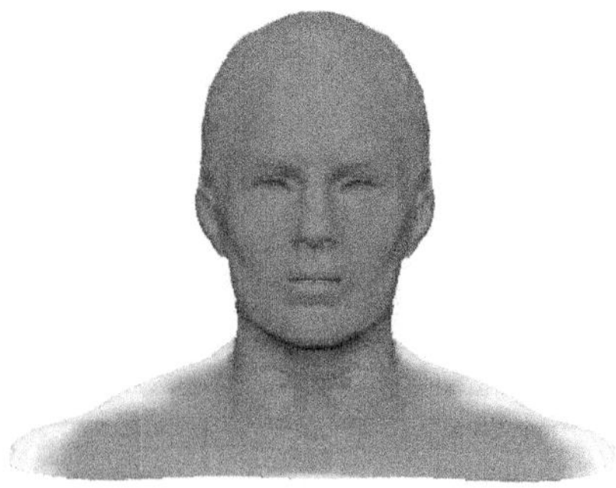

Figure 10-8 *The front side of the model painted using the* ***Protect*** *button*

6. In the **UV Master** subpalette, choose the **Attract** button and paint on the back side of the model to display the seams on it, refer to Figure 10-9.

7. Choose the **Unwrap** button; the seam is displayed on the model, as shown in Figure 10-10.

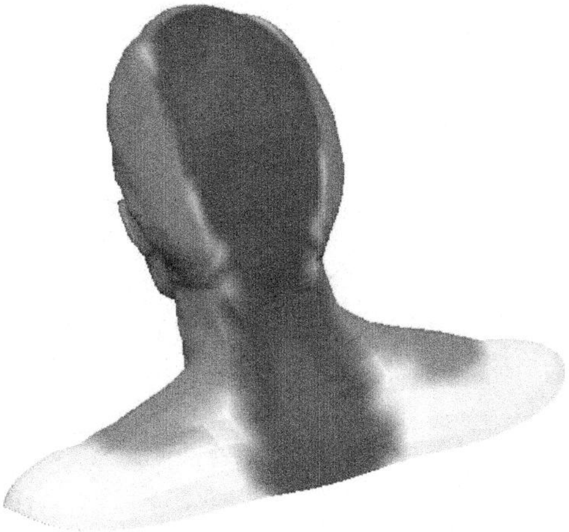

Figure 10-9 *The back side of the model painted using the* ***Attract*** *button*

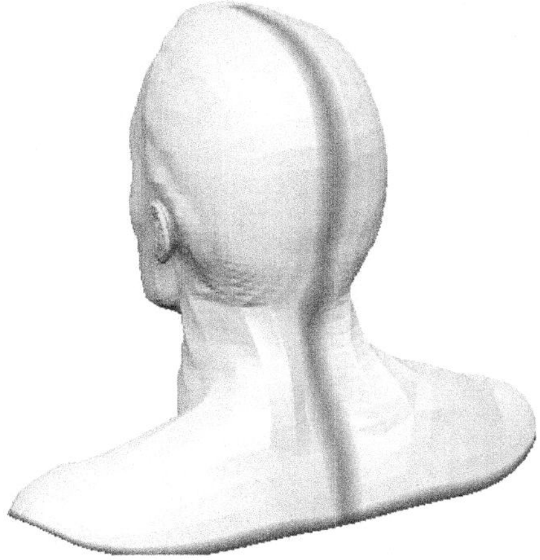

Figure 10-10 *The seam displayed after choosing the* ***Unwrap*** *button*

8. Choose the **Flatten** button; the flattened image of the model is displayed, refer to Figure 10-11. In this figure, you will notice that the size of UVs on the face area is too small. Choose the **UnFlatten** button to switch back to the 3D model.

   Next, you will increase the size of the UVs on the face, so that the texture is applied properly.

9. Choose the **Density** button in the **UV Master** subpalette. Next, set the value in the **density** slider to **4**, refer to Figure 10-12.

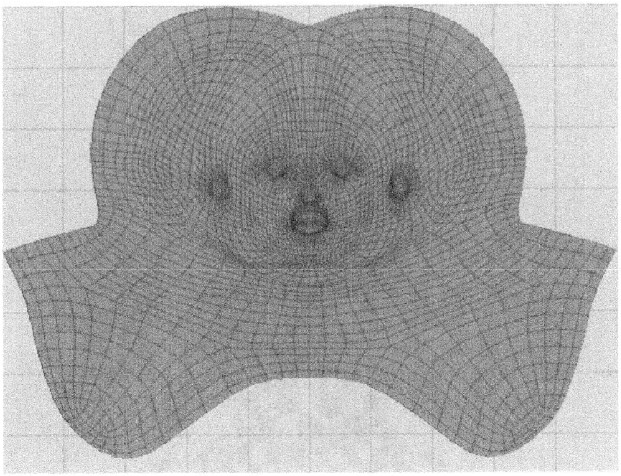

Figure 10-11 *The flattened image of the model displayed*

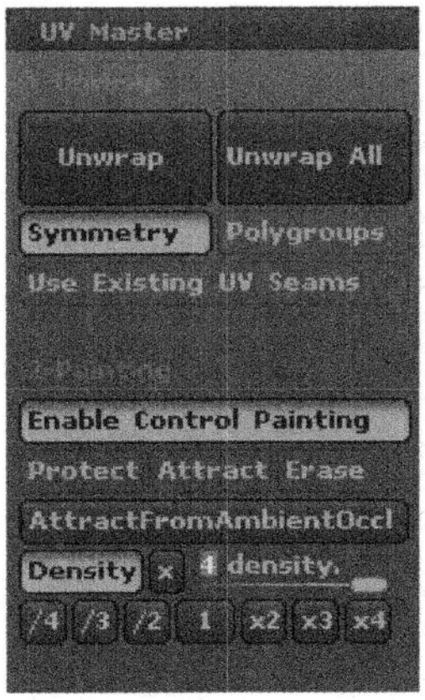

Figure 10-12 *The value of the* ***density*** *slider set to* ***4***

10. With the **Density** button chosen, start painting on the face area of the model, refer to Figure 10-13.

11. Choose the **Unwrap** button in the **UV Master** subpalette. Next, choose the **Flatten** button; the flattened image of the model is displayed, as shown in Figure 10-14.

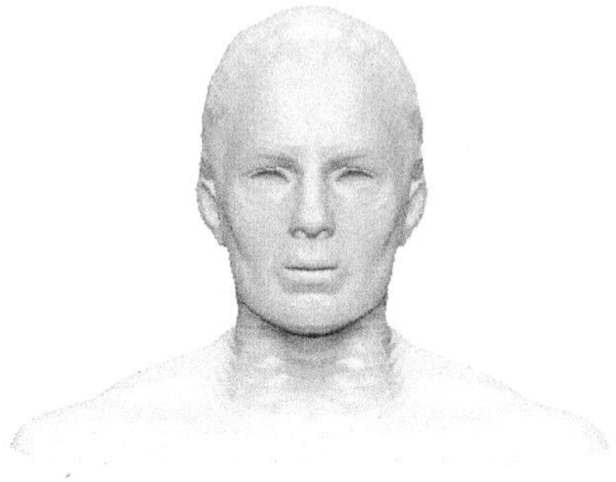

Figure 10-13 *The model with face area painted*

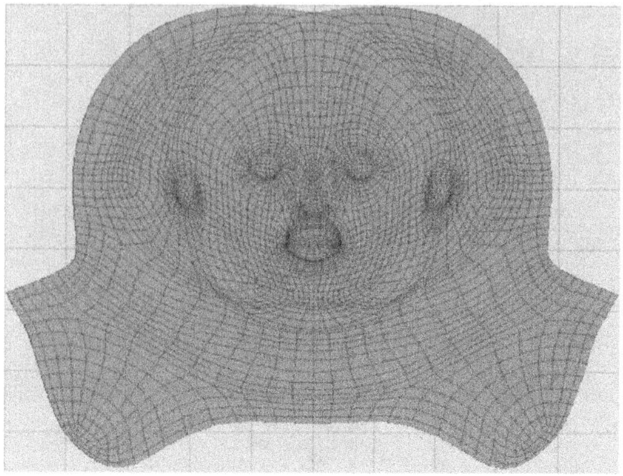

Figure 10-14 *The flattened image of the unwrapped model*

12. Choose the **UnFlatten** button to switch back to the 3D model. Next, choose the **Copy UVs** button, refer to Figure 10-15.

13. Choose the Current Tool button in the **Tool** palette; a flyout is displayed. In this flyout, choose the original *DemoHead* model; the model is displayed in the canvas.

14. Choose the **Paste UVs** button in the **UV Master** subpalette; the UVs are pasted on the original model, refer to Figure 10-16.

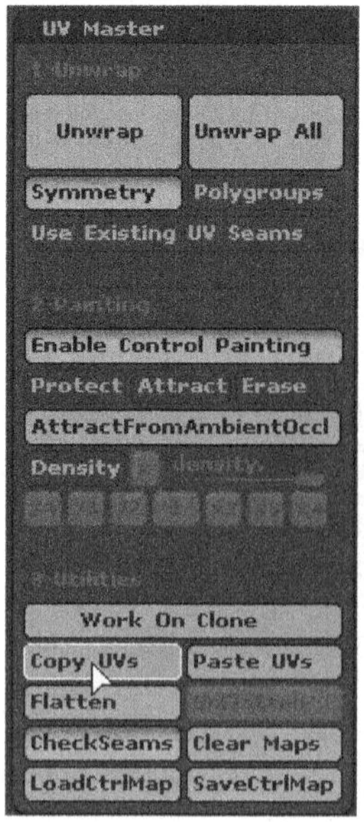

Figure 10-15 *The* ***Copy UVs*** *button chosen*

Figure 10-16 *The UVs pasted on the original model*

Saving the Model

In this section, you will save the file using the steps given next.

1. Choose the **Save As** button from the **Tool** palette, the **Save ZTool** dialog box is displayed. In this dialog box, browse to the location *\Documents\ZBrushprojects\c10*.

2. Enter **c10tut1** in the **File name** edit box and then choose the **Save** button.

Tutorial 2

In this tutorial, you will create a texture map and apply it to a *Dog.ZTL* model. The final output of the model is shown in Figure 10-17. **(Expected time: 20 min)**

Figure 10-17 *Texture map applied to the model*

The following steps are required to complete this tutorial:

a. Unwrap the model.
b. Paint on the flattened image.
c. Create a texture map and apply it to the model.
d. Save the model.

Unwrapping the Model

In this section, you will unwrap the model using the default settings of the **UV Master** subpalette.

1. Choose the **Init ZBrush** button from the **Preferences** palette; the message box is displayed. Choose the **Yes** button from the message box; ZBrush is initialized to its default state.

2. Choose the **Dog.ZTL** file from the **Tool** tab of the LightBox browser by double-clicking on it. Next, drag the cursor in the canvas area; the *Dog.ZTL* is displayed in the canvas, refer to Figure 10-18.

3. Choose the **Edit** button in the top shelf to switch to the edit mode.

4. Dock the **ZPlugin** palette in the right tray. In this palette, expand the **UV Master** subpalette. In this subpalette, choose the **Unwrap** button.

5. Choose the **CheckSeams** button in the **UV Master** subpalette; the seams are displayed on the model, refer to Figure 10-19.

Figure 10-18 *The Dog.ZTL model displayed in the canvas*

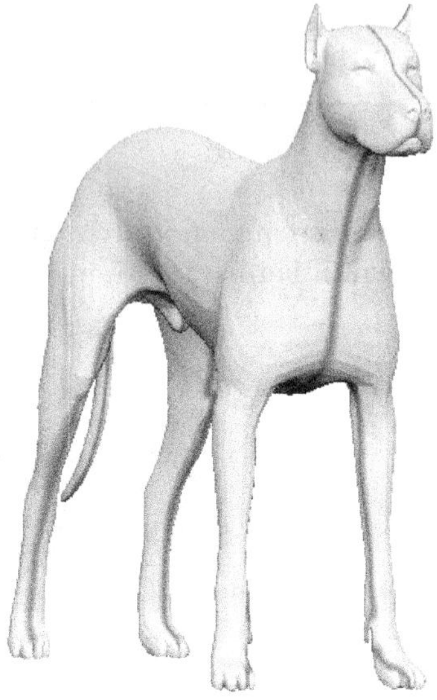

Figure 10-19 *The seams displayed on the model*

6. Choose the **Flatten** button in the **UV Master** subpalette; the flattened image of the model is displayed, as shown in Figure 10-20. For painting on the image, you need to increase its subdivisions level. To do so, press CTRL+D thrice; the flattened image is displayed, as shown in Figure 10-21.

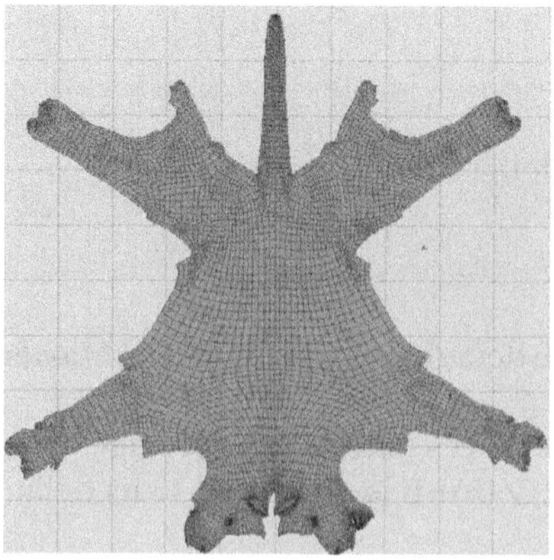

Figure 10-20 *The flattened image of the model*

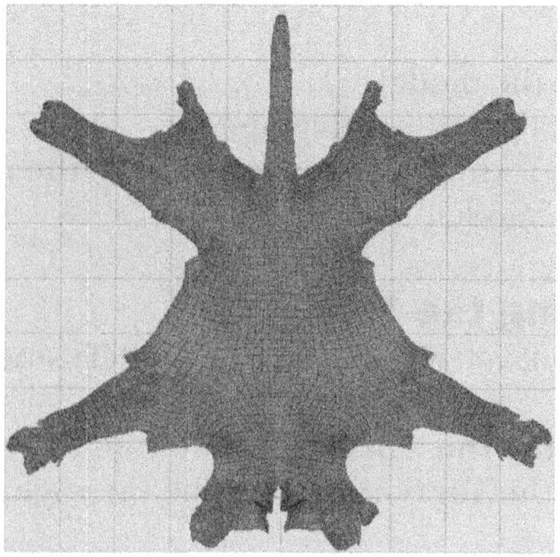

Figure 10-21 *The flattened image after pressing CTRL+D thrice*

Painting on the Flattened Image

In this section, you will paint on the surface of the flattened image.

1. Choose the Current Material button from the left shelf; a flyout is displayed. In this flyout, choose the **SkinShade4** material from the **Standard Materials** area; the material is applied on the flattened image.

2. To view the material, choose the **PolyF** button in the right shelf; the flattened image is displayed, as shown in Figure 10-22.

3. Expand the **Color** palette. In this palette, choose the **FillObject** button to fill the image with white color.

4. Choose the **PolyF** button again to display the polygon edges on the flattened image.

5. Make sure the **Rgb** button is chosen and the **Zadd** button is not chosen in the top shelf. Choose the black color from the **Current Color** swatch. Next, paint on the surface of the image to create the patterns, as shown in Figure 10-23.

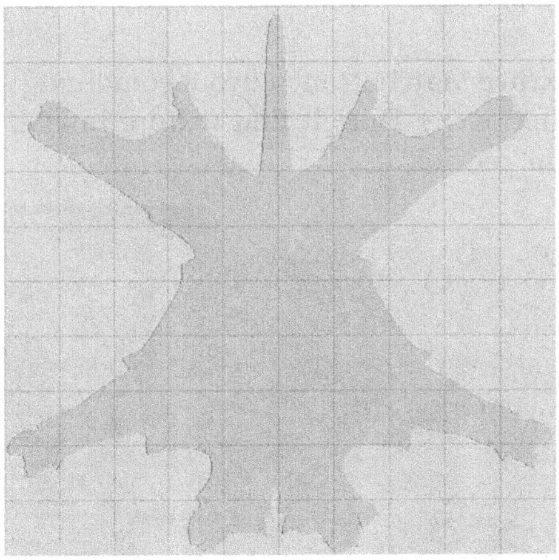

***Figure 10-22** The **SkinShade 4** material applied to the image*

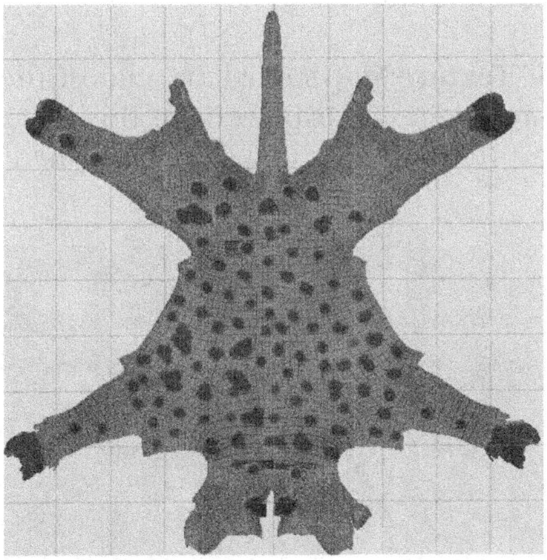

***Figure 10-23** A pattern created on the surface of the image*

Creating a Texture Map and Applying it to the Model

In this section, you will create a texture map from the flattened image and then apply it to the model.

1. Expand the **Texture Map** subpalette in the **Tool** palette. In this subpalette, choose the **Create** option; it expands. Next, choose the **New From Polypaint** button, refer to Figure 10-24; the flattened image is displayed in the **Texture Map** button, refer to Figure 10-25.

2. In the **Texture Map** subpalette, choose the **Clone Txtr** button, refer to Figure 10-25; the flattened image is cloned and is added to the **Texture** palette.

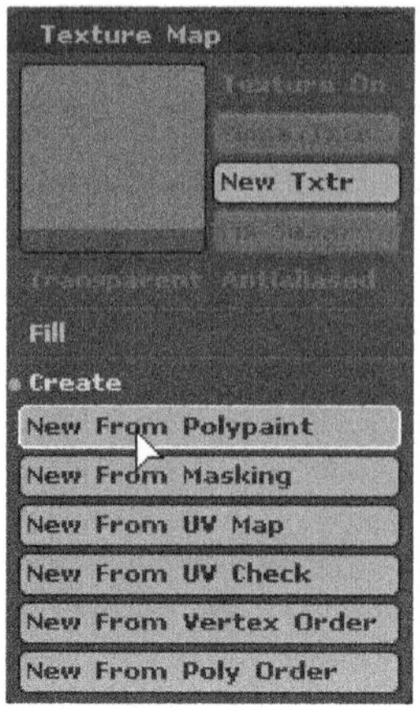

Figure 10-24 *The* ***New From Polypaint*** *button chosen*

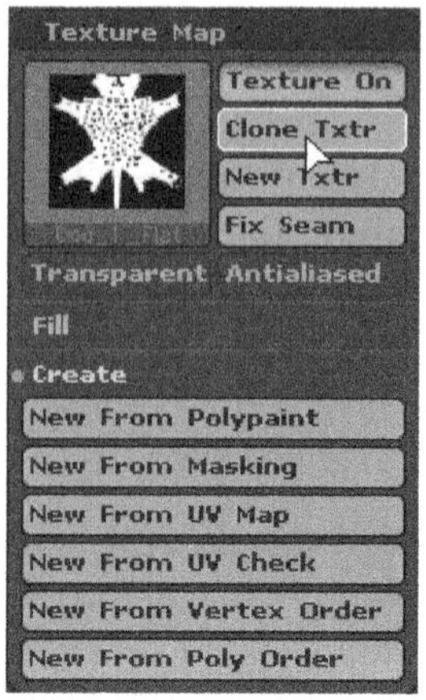

Figure 10-25 *The* ***Clone Txtr*** *button chosen*

3. In the **UV Master** subpalette, choose the **UnFlatten** button; the 3D model of the dog is displayed in the canvas.

4. In the **Texture Map** subpalette, choose the **Texture Map** button; a flyout containing different texture images is displayed. In this flyout, choose the **Dog_1_Flat** texture image, refer to Figure 10-26; the texture map is applied to the model.

Figure 10-26 *The texture image chosen from the flyout*

Saving the Model

In this section, you will save the file using the steps given next.

1. Choose the **Save As** button from the **Tool** palette, the **Save ZTool** dialog box is displayed. In this dialog box, browse to the location *\Documents\ZBrushprojects\c10*.

2. Enter **c10tut2** in the **File name** edit box and then choose the **Save** button.

EXERCISE

The output of the model used in this exercise can be accessed by downloading the *c10_ZBrush_2018_exr.zip* from *www.cadsofttech.com*. The path of the file is as follows: *Textbooks > Animation and Visual Effects > Pixologic ZBrush 2018 > Pixologic ZBrush 2018 for Novices.*

Exercise 1

Load the *DemoHead.ZTL* file from the **Tool** tab of the LightBox browser and then unwrap it. Next, create a texture map and apply it to the model, as shown in Figure 10-27.

(Expected time: 20 min)

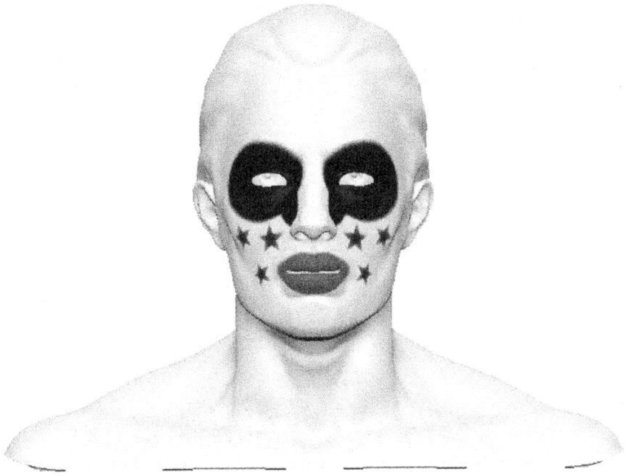

Figure 10-27 *The final output*

Page intentionally left blank

Chapter 11

Lighting

Learning Objectives

After completing this chapter, you will be able to:

- *Work with the Light palette*
- *Understand the different types of lights in ZBrush*
- *Light up a scene using the three point lighting setup*

INTRODUCTION

In this chapter, you will work with different lights in ZBrush. Lighting is used to illuminate a scene. The **Light** palette in ZBrush is used to add lights to a scene and then to manipulate them using different settings.

Light PALETTE

The **Light** palette has different settings and subpalettes that are used to manipulate the lights to produce the desired outputs, refer to Figure 11-1. Before applying the lights to a scene, you have to make sure that MatCap materials are not applied to the objects in a scene. The MatCap materials do not respond to the lights in a scene as they already have lights baked into them. Therefore, you need to apply the Standard materials to the objects in a scene before adding lights. The different settings and subpalettes in the **Light** palette are discussed next.

Load and Save

The **Save** button is used to save the different settings of a light. The **Load** button is used to load the saved light settings into a different scene. To load the saved light settings into a different scene, create different objects in the canvas, refer to Figure 11-2. Next, apply the **BasicMaterial** material to them.

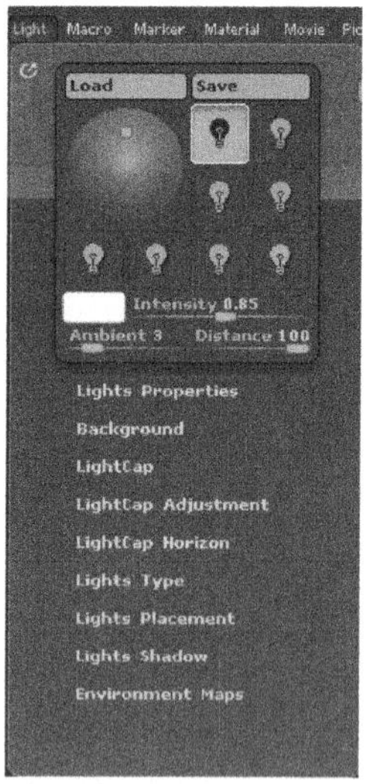

*Figure 11-1 The **Light** palette*

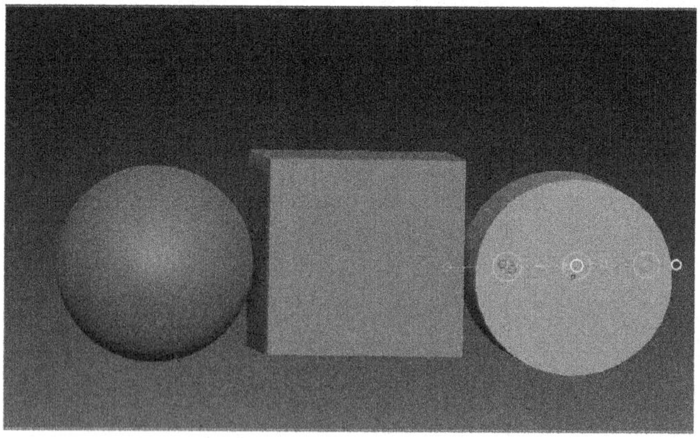

Figure 11-2 Different objects created in the canvas

Expand the **Light** palette. In this palette, change the color of the light to yellow by choosing the Light Color swatch, refer to Figure 11-3. Set the value of the **Intensity** and **Ambient** sliders to **6.6183** and **44**, respectively; the light settings will be applied to the scene, refer to Figure 11-4. These settings will be discussed in detail later in this chapter.

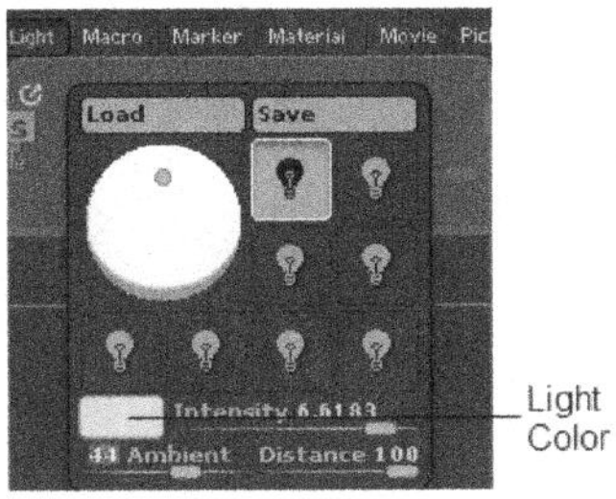

*Figure 11-3 Different settings in the **Light** palette*

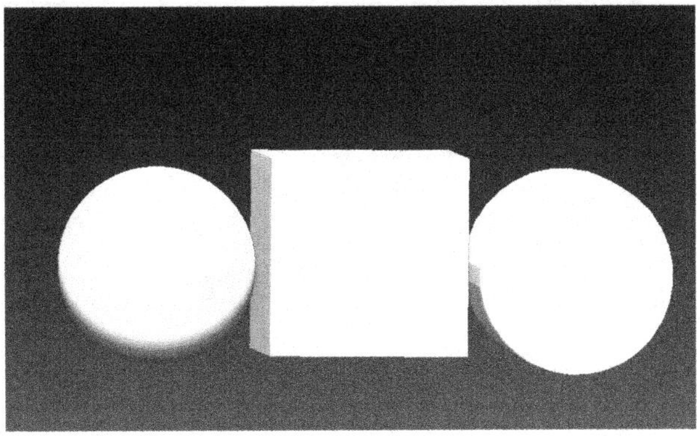

Figure 11-4 Light settings applied to the scene

Choose the **Save** button in the **Light** palette; the **Save Lights** dialog box will be displayed. In this dialog box, choose the **Save** button; the light settings will be saved with the name **ZLights.zli**, refer to Figure 11-5.

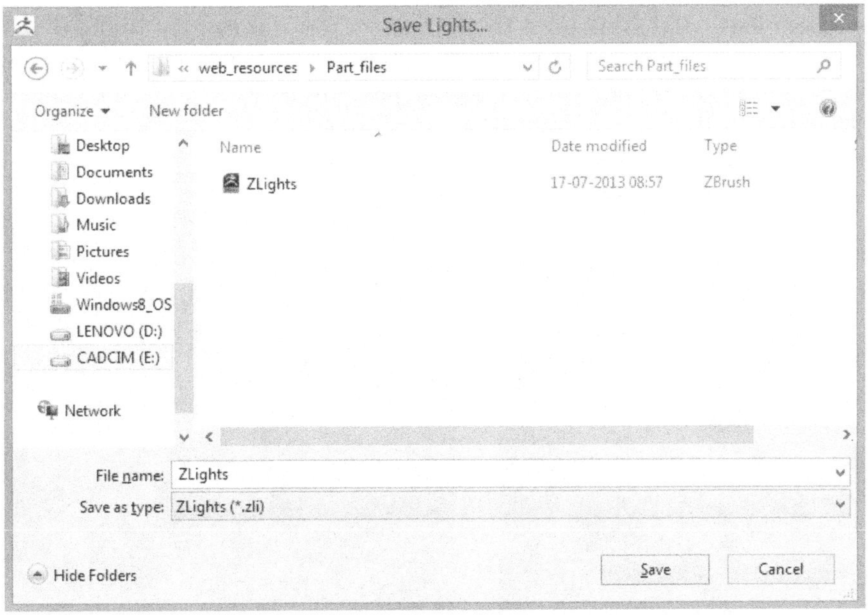

*Figure 11-5 The **Save Lights** dialog box*

Choose the **Init ZBrush** button from the **Preferences** palette; ZBrush is initialized to its default state. Next, create an object in the canvas and assign the **BasicMaterial** material to it, refer to Figure 11-6. Expand the **Light** palette. In this palette, choose the **Load** button; the **Load Lights** dialog box will be displayed. In this dialog box, choose the **Open** button; the light settings applied to the previous scene will be applied to the new scene also, refer to Figure 11-7.

*Figure 11-6 The **BasicMaterial** material applied to the object*

Figure 11-7 The light settings applied in the new scene

To save the scene with lights, you need to save the scene using the **File** palette. To do so, expand the **File** palette. In this palette, choose the **Save As** button; the **Save Project** dialog box will be displayed. In this dialog box, enter the desired name in the **File Name** text box and then choose the **Save** button. If you save the scene using the **Save As** button in the **Tool** palette, only the 3D scene will be saved and the lighting information will be lost.

Light Placement

The **Light Placement** window is used to change the position of the light in a scene. The orange dot in this window represents the position of the light. On moving this orange dot, the position of the light in the scene changes accordingly and is reflected in the canvas area, refer to Figures 11-8 and 11-9.

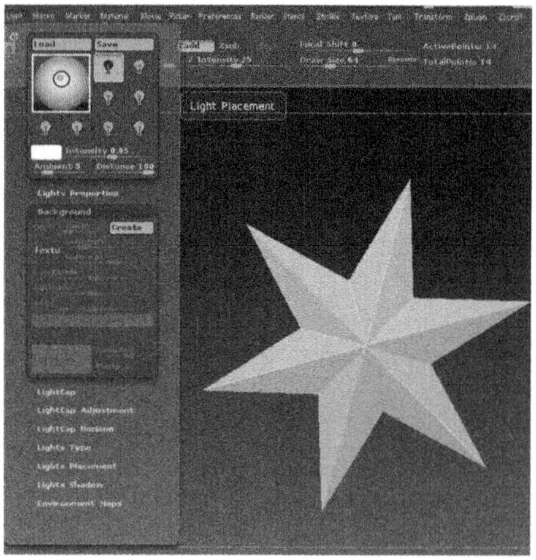

*Figure 11-8 The position of light in the **Light Placement** window*

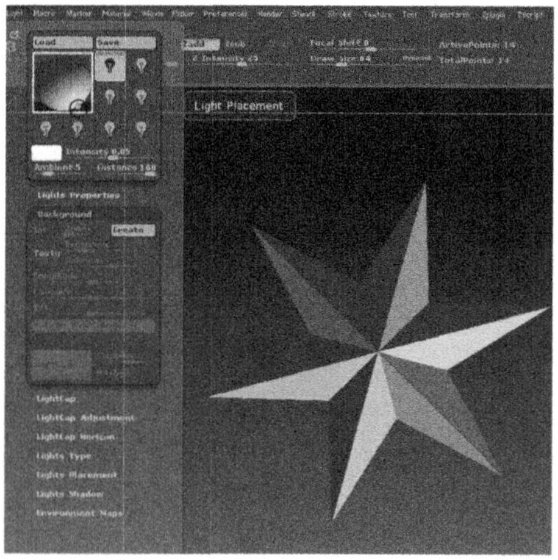

*Figure 11-9 The position of light changed in the **Light Placement** window*

Light Switches

The Light Switches are used to switch on or switch off the lights in a scene. There are eight light switches in the **Light** palette, refer to Figure 11-10. By default, only one light switch is switched on. To switch on the other lights, you need to select the light switch and then click on the bulb icon. The light that is switched on will be highlighted in the palette.

Each light has different intensity. You can change the intensity, ambient value, and the color of each light by selecting it and then setting the values as required. Figure 11-11 displays a scene with all the lights turned on.

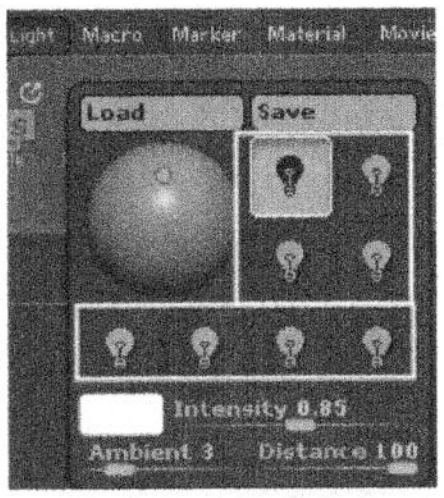

Figure 11-10 The light switches in the ***Light*** *palette*

Figure 11-11 The scene illuminated by switching on all the lights

Light Color

The Light Color swatch is used to change and display the color of the selected light. If you set different colors for each light, all the colors will be blended together in the scene.

Intensity

The **Intensity** slider is used to increase or decrease the brightness of a scene. The greater the value of this slider, the higher will be the brightness in the scene. Figures 11-12 and 11-13 display a scene with the value of the **Intensity** slider set to **0** and **10**, respectively.

Figure 11-12 The scene with the value of the ***Intensity*** *slider set to* ***0***

*Figure 11-13 The scene with the value of the **Intensity** slider set to **10***

Ambient

The **Ambient** slider is used to set the value of brightness in an object even if no light is present in the scene. To understand the working of this slider, turn off all the lights in the **Light** palette, refer to Figure 11-14. Next, set the value of the **Ambient** slider to **100**; the objects in the scene will be lit up without any source of light, refer to Figure 11-15.

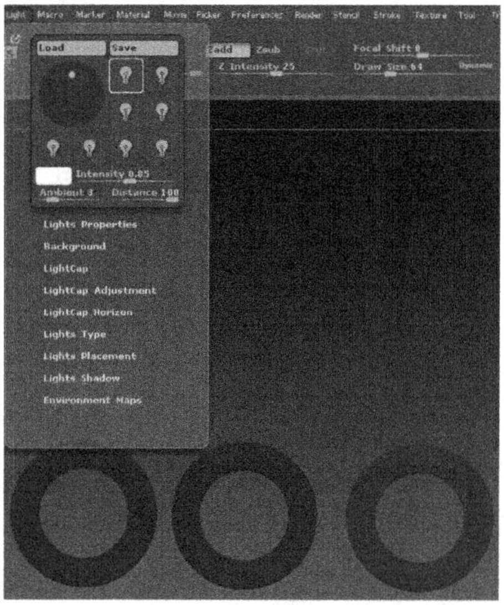

Figure 11-14 All the lights turned off

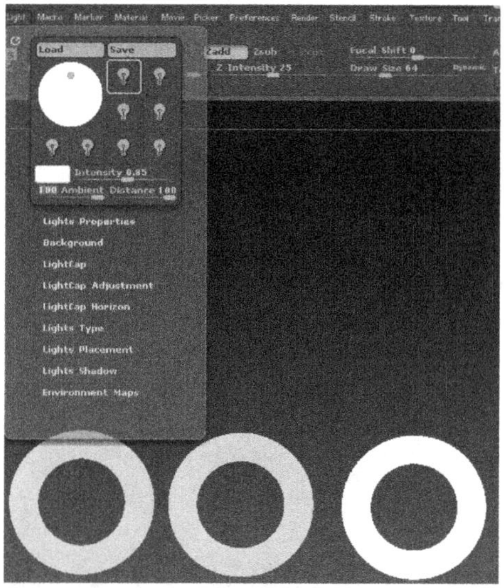

*Figure 11-15 The value of the **Ambient** slider set to **100***

Distance

The **Distance** slider is used to adjust the distance of lights from the objects in a scene. By default, the value of this slider is set to 100 which means that the light is at an infinite distance from the objects in the scene.

Lights Properties

The **Lights Properties** subpalette consists of different attributes that are used to control the shadows, the Subsurface scattering, and the intensity of the light in the scene.

Background

The **Background** subpalette consists of different attributes that are used to create and manipulate a background image in the scene.

LightCap

The **LightCap** subpalette consists of attributes that are used to create a light capture using different lights and adjust different settings that can be saved and then loaded into some other scene. This subpalette consists of different buttons and sliders, refer to Figure 11-16.

***Figure 11-16** The **LightCap** subpalette*

LightCap Adjustment

The **LightCap Adjustment** subpalette is used to globally adjust the settings of the entire light capture and not of the single lights in the light capture. The different settings in the **LightCap** subpalette enable you to manipulate each light in the light capture individually by selecting a particular light and applying the settings to it. By default, the different settings in the **LightCap Adjustment** subpalette are deactivated. These settings are activated only if a light capture is created using the **LightCap** subpalette.

Lights Placement

The **Lights Placement** subpalette is used to change the position of the light in a scene. However, the settings in this subpalette do not work for the **Sun** light. This subpalette consists of a picker button **P**, refer to Figure 11-17. This picker button is also known as Local Light Position Selector button. The picker button enables you to pick the position of the point light, spot light, or glow light from the canvas. On picking the position of required light from the canvas, the value of the **X Pos**, **Y Pos**, and **Z Pos** sliders will change according to the picked position of the light. To pick the position of the light from the canvas, choose the required light from the **Lights Type** subpalette. Make sure, you do not select the **Sun** light.

Next, click on the **P** button in the **Lights Placement** subpalette and then drag the cursor on the canvas; the shape of the cursor will change and the values in the **X Pos**, **Y Pos**, and **Z Pos** sliders will also change, refer to Figure 11-18. The **Radius** slider is used to specify the radius of the selected light.

Lights Shadow

The **Lights Shadow** subpalette is used to control the shadows that are cast by the selected light.

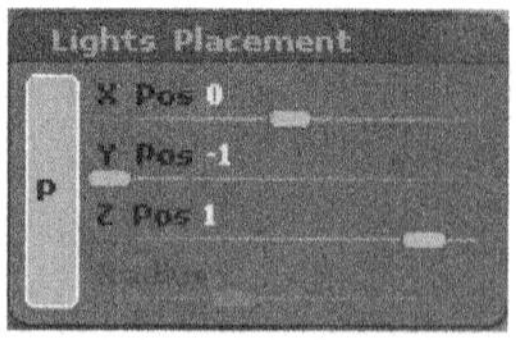

*Figure 11-17 The **Lights Placement** subpalette*

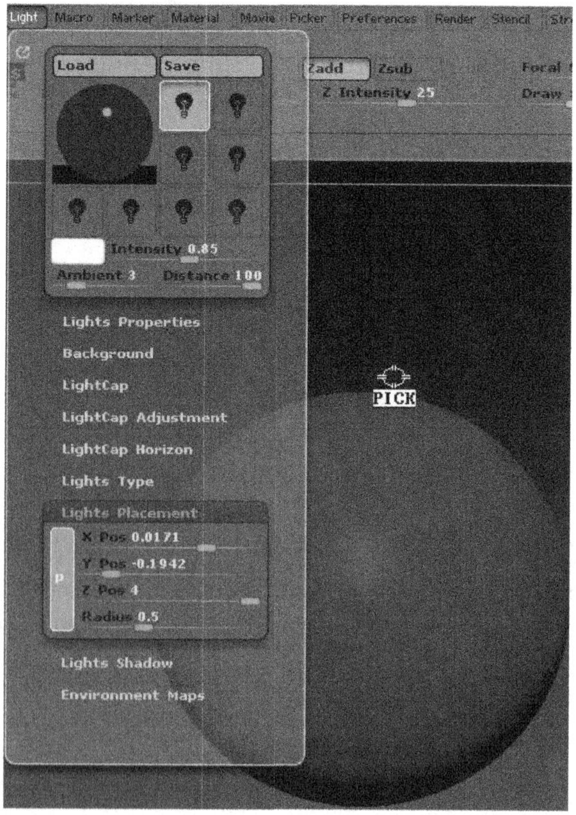

*Figure 11-18 The values of the **X Pos**, **Y Pos**, and **Z Pos** sliders updated*

Environment Maps

The **Environment Maps** subpalette is used to control the global illumination of the environment in a scene.

TUTORIAL

All the files used in the tutorial can be downloaded from the CADSoft website (*www.cadsofttech.com*).These files are compressed in zip file format and are required to be extracted before using them in the tutorials. The path of the files is as follows: *Textbooks > Animation and Visual Effects > Pixologic ZBrush 2018 > Pixologic ZBrush 2018 for Novices*

Tutorial 1

In this tutorial, you will apply three point lighting to a model. The final output of the lighting is shown in Figure 11-19. **(Expected time: 20 min)**

The following steps are required to complete this tutorial:

a. Create the key light.
b. Create the fill light.
c. Create the back light and render the scene.
d. Save the scene.

Creating the Key Light

In this section, you will create a key light. The key light is the main source of illumination in the scene.

1. Choose the **Init ZBrush** button from the **Preferences** palette; the message box is displayed. Choose the **Yes** button from the message box; ZBrush is initialized to its default state.

2. Choose the **Ryan_Kingslien_Anatomy_Model.ZTL** file from the **Tool** tab of the **LightBox** browser by double-clicking on it. Next, drag the cursor in the canvas area; the model is displayed in the canvas.

3. Choose the **Edit** button in the top shelf to switch to the edit mode.

4. Choose the Current Material button from the left shelf; a flyout consisting of different types of materials is displayed. In this flyout, choose the **FastShader** material; the material is applied to the model, refer to Figure 11-20.

   In order to access the **Light** palette easily, you need to dock the **Light** palette in the right or left tray.

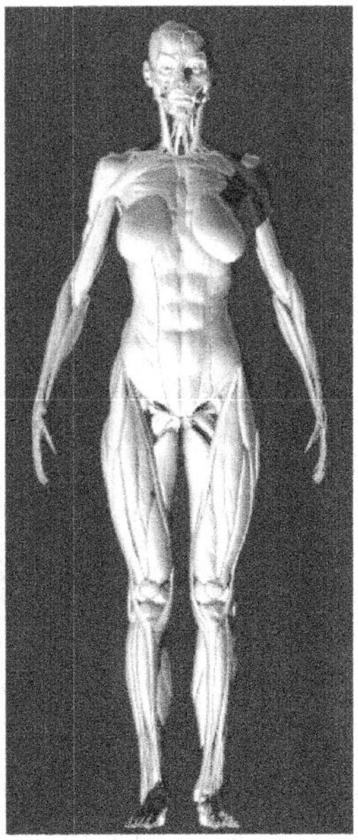

Figure 11-19 *The output after applying lights*

5. Expand the **Light** palette and hover the cursor on the arrow located at the top left corner of the palette; the shape of cursor changes, refer to Figure 11-21. Next, drag the cursor toward the right tray; the **Light** palette is docked in the right tray and can be easily accessed.

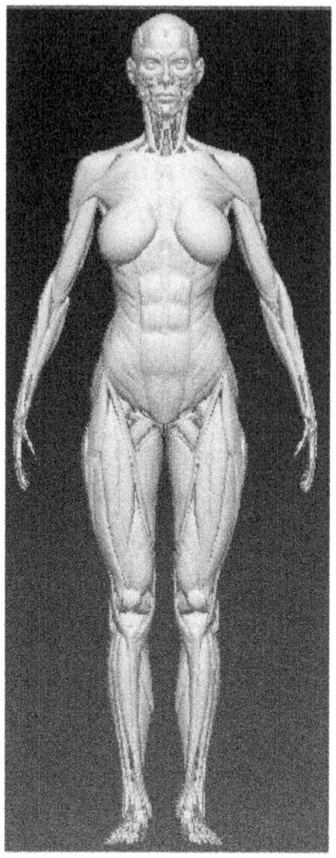

Figure 11-20 *The* ***FastShader*** *material applied to the model*

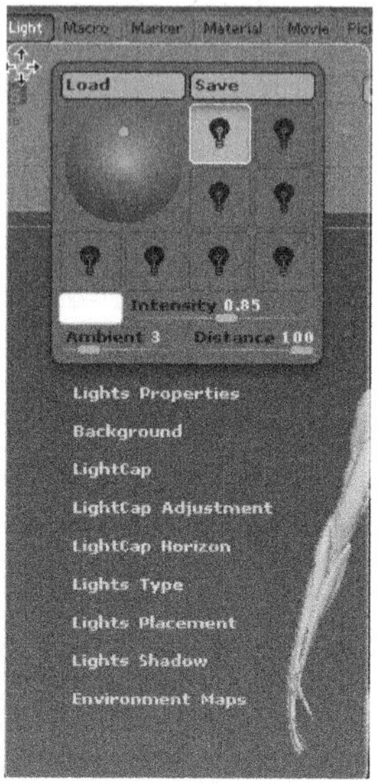

Figure 11-21 *The* ***Light*** *palette*

6. In the **Light** palette, make sure the first light switch is turned on.

7. Select the orange dot on the **Light Placement** window and drag it to the location, as shown in Figure 11-22.

8. Set the **Intensity** slider to **1**, refer to Figure 11-22.

9. In the **Light** palette, expand the **Light Properties** subpalette and choose the **Shadow** button, if not already chosen, refer to Figure 11-23.

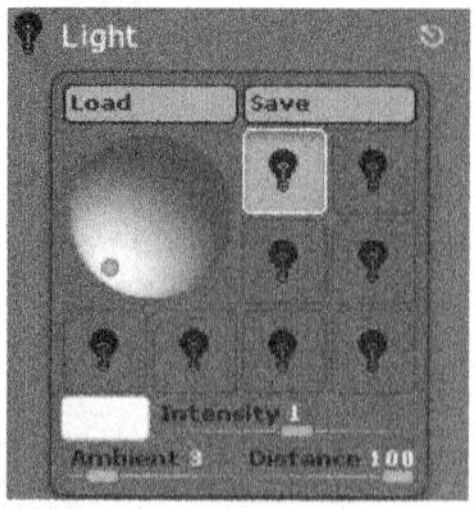

Figure 11-22 *The key light positioned in the* ***Light Placement*** *window*

Figure 11-23 *The* ***Shadow*** *button chosen*

10. Expand the **Lights Type** subpalette in the **Light** palette. In this subpalette, choose the **Sun** button to activate the sunlight, if not already chosen, refer to Figure 11-24; the key light is created.

Creating the Fill Light

In this section, you will create a fill light. The fill light is used to fill the shadows generated by the key light to make the object more visible.

1. Double-click on the second light switch to activate the second light. Make sure the light switch turns orange when you click on it.

2. Select the orange dot on the **Light Placement** window and drag it to the position, as shown in Figure 11-50.

3. Choose the Light Color swatch located below the light switches and choose light red shade from the color picker. Next, set the value of the **Intensity** slider to **0.5**, refer to Figure 11-25.

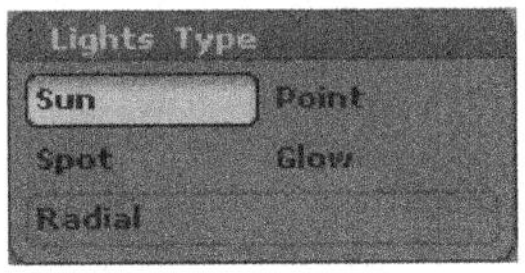

*Figure 11-24 The **Sun** button chosen*

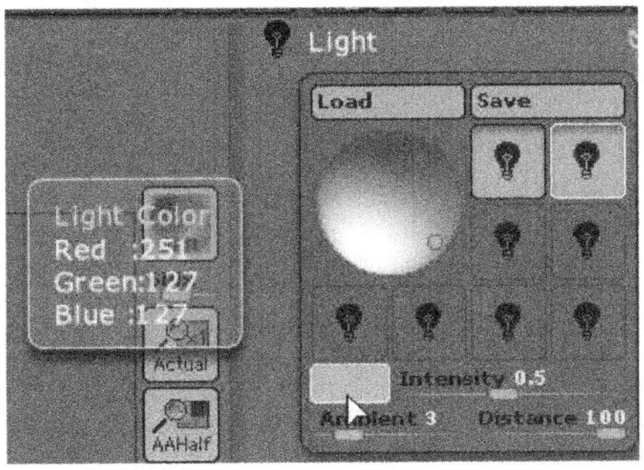

*Figure 11-25 The fill light positioned in the **Light Placement** window with its color changed*

4. Expand the **Lights Type** subpalette in the **Light** palette. In this subpalette, choose the **Spot** button to activate the spot light, refer to Figure 11-26; the fill light is created.

Creating the Back Light and Rendering the Scene

In this section, you will create a back light. The back light is used to separate the object in the scene from the background.

1. Double-click on the third light switch to activate the third light. Make sure the light switch turns orange.

2. Select the orange dot on the **Light Placement** window and drag it to the position, as shown in Figure 11-27.

3. Set the value of the **Intensity** slider to **0.3**, refer to Figure 11-27.

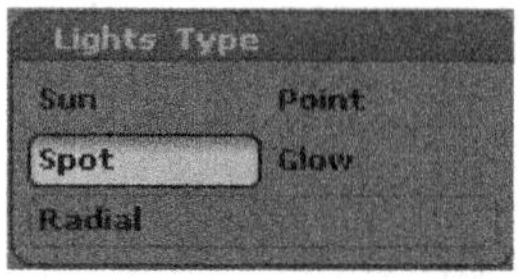

*Figure 11-26 The **Spot** button chosen*

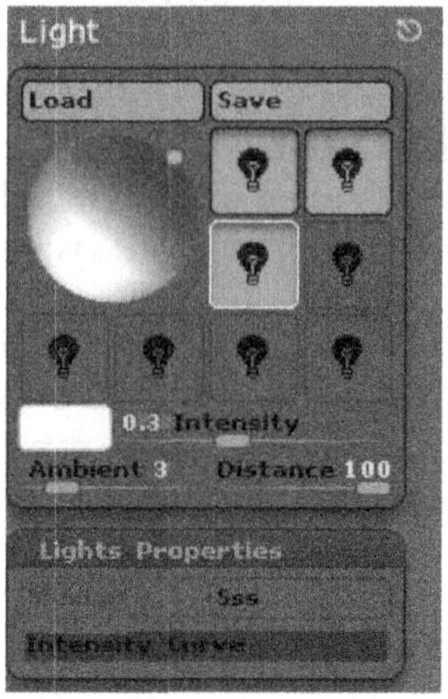

*Figure 11-27 The back light positioned in the **Light Placement** window*

4. Expand the **Lights Type** subpalette in the **Light** palette. In this subpalette, choose the **Spot** button to activate the spot light; the back light is created.

5. Expand the **Background** subpalette in the **Light** palette. In this subpalette, choose the **Create** button to create the panoramic background, refer to Figure 11-28.

6. Choose the **BPR** button in the right shelf; the model is rendered, as shown in Figure 11-29.

Saving the File

In this section, you will save the file using the steps given next.

1. Choose the **Save As** button from the **File** palette, the **Save Project** dialog box is displayed. In this dialog box, browse to the location *\Documents\ZBrushprojects\c11*.

2. Enter **c11tut1** in the **File name** edit box and then choose the **Save** button.

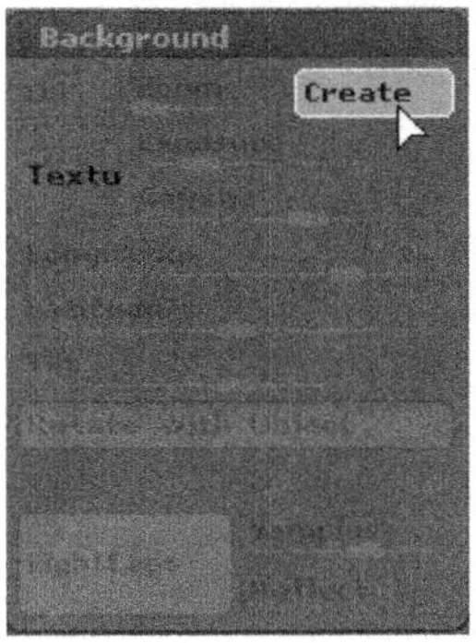

Figure 11-28 *The* ***Create*** *button chosen in the* ***Background*** *subpalette*

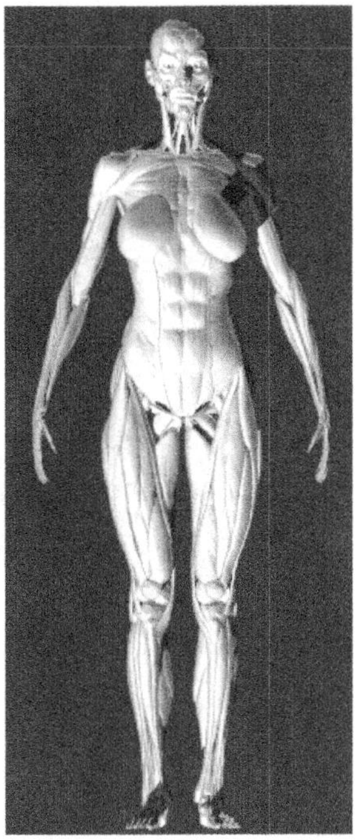

Figure 11-29 *The model rendered using the* ***BPR*** *renderer*

EXERCISE

The output of the model used in this exercise can be accessed by downloading the *c11_ZBrush_2018_exr.zip* from *www.cadsofttech.com*. The path of the file is as follows: *Textbooks > Animation and Visual Effects > Pixologic ZBrush 2018 > Pixologic ZBrush 2018 for Novices.*

Exercise 1

Load the *DemoHead.ZTL* file from the **Tool** tab of the **LightBox** browser and then light it up using the three point lighting set up, as shown in Figure 11-30. **(Expected time: 20 min)**

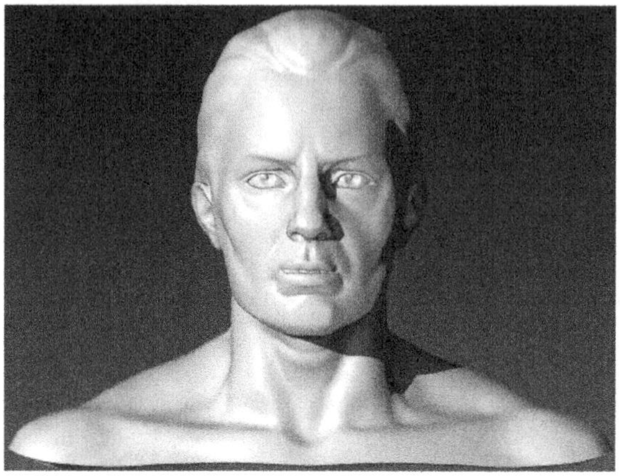

Figure 11-30 *The final output*

Chapter 12

Rendering

Learning Objectives

After completing this chapter, you will be able to:

- *Work with the Render palette*
- *Generate render passes for compositing*

INTRODUCTION

In this chapter, you will work with different render settings in ZBrush. Rendering is a process of generating a 2D image for a 3D or 2.5D scene created in ZBrush. Rendering helps in visualizing the lighting effects, materials, and other settings that have been applied to a scene. ZBrush provides several ways to render both 2D and 3D scene.

TUTORIAL

All the files used in the tutorial can be downloaded from the CADSoft website (*www.cadsofttech.com*). These files are compressed in zip file format and are required to be extracted before using them in the tutorials. The path of the files is as follows: *Textbooks > Animation and Visual Effects > Pixologic ZBrush 2018 > Pixologic ZBrush 2018 for Novices.*

Tutorial 1

In this tutorial, you will set up a scene and create render passes of the scene.

(Expected time: 20 min)

The following steps are required to complete this tutorial:

a. Set up a scene.
b. Generate different render passes.
c. Save the scene.

Setting up the Scene

In this section, you will set up a scene.

1. Choose the **Init ZBrush** button from the **Preferences** palette; ZBrush is initialized to its default state.

2. Choose the **DemoHead.ZTL** file from the **Tool** tab of the **LightBox** browser by double-clicking on it. Next, choose the **LightBox** button to close the **LightBox** browser. Now, drag the cursor in the canvas area; the model is displayed in the canvas.

3. Choose the **Edit** button in the top shelf to switch to the edit mode.

4. Expand the **Geometry** subpalette in the **Tool** palette. In this subpalette, click twice on the **Divide** button; the value of **SDiv** slider becomes equal to **5**.

5. Expand the **SubTool** subpalette in the **Tool** palette. In this subpalette, choose the **Append**

button; a flyout is displayed. In this flyout, choose the **Plane3D** primitive, refer to Figure 12-1. On doing so, the plane is displayed in the canvas, as shown in Figure 12-2.

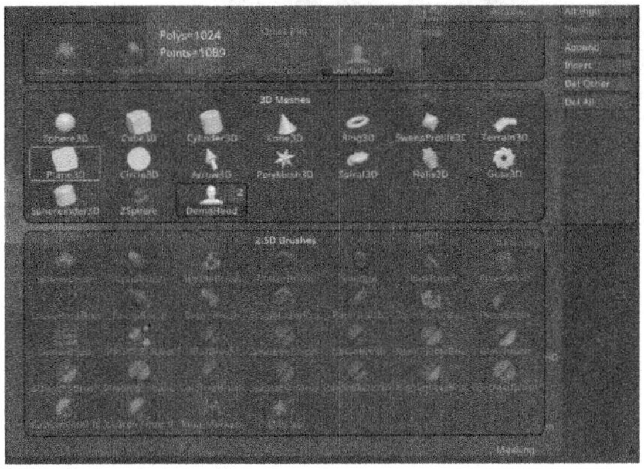

*Figure 12-1 The **Plane3D** primitive chosen from the flyout*

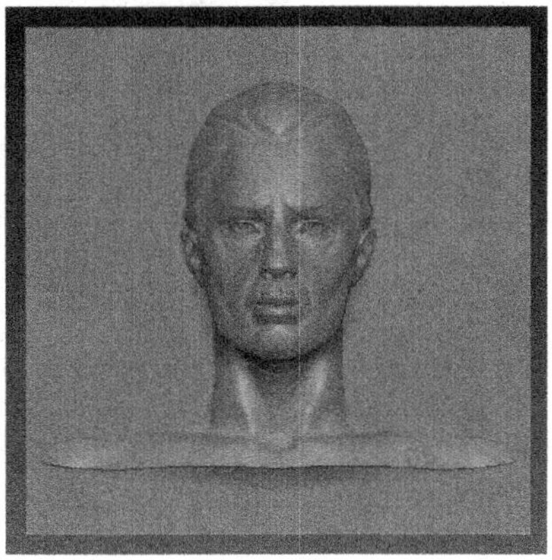

Figure 12-2 The plane displayed in the canvas

6. Make sure the plane is selected in the SubTool list in the **SubTool** subpalette of the **Tool** palette. Expand the **Deformation** subpalette in the **Tool** palette.

   You need to rotate the plane along the X-axis using the **Rotate** slider in the **Deformation** subpalette.

7. Deactivate the **y** and **z** buttons corresponding to the **Rotate** slider and activate the **x** button. Next, set the value of the **Rotate** slider to **90**, refer to Figure 12-3; the plane rotates in the canvas, however it is not visible in the canvas.

8. Expand the **Display Properties** subpalette in the **Tool** palette. In this subpalette, make sure the **Double** button is chosen, refer to Figure 12-4; the rotated plane is visible in the canvas.

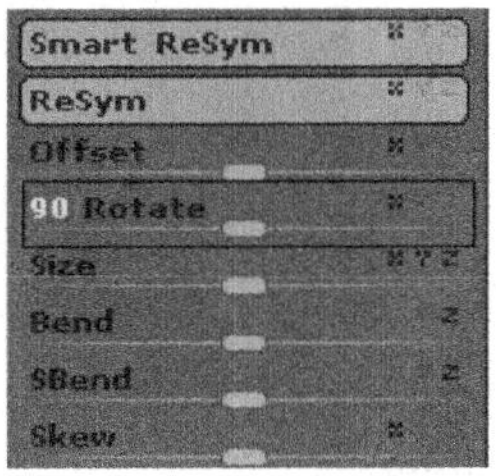

*Figure 12-3 The value of the **Rotate** slider set to **90***

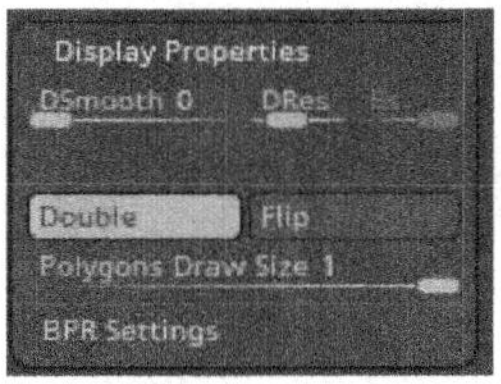

*Figure 12-4 The **Double** button chosen*

9. Increase the size of the plane using the **Size** slider in the **Deformation** subpalette. Next, move the plane below the model using the **Move** tool, refer to Figure 12-5. Now, the scene is set.

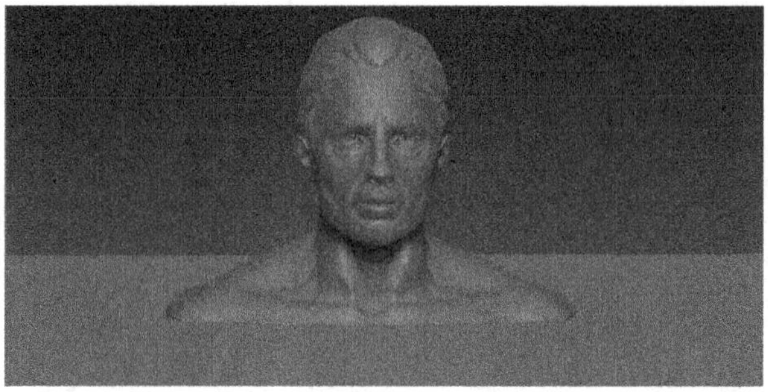

Figure 12-5 *The plane placed below the model*

Generating Different Render Passes

In this section, you will generate different render passes of the scene using the **BPR** renderer.

1. Expand the **Light** palette. In this palette, make sure the first light switch is turned on.

2. Select the orange dot on the **Light Placement** window and drag it to the position shown in Figure 12-6.

3. Expand the **Render Properties** subpalette in the **Render** palette. In this subpalette, choose the **AOcclusion** button, refer to Figure 12-7.

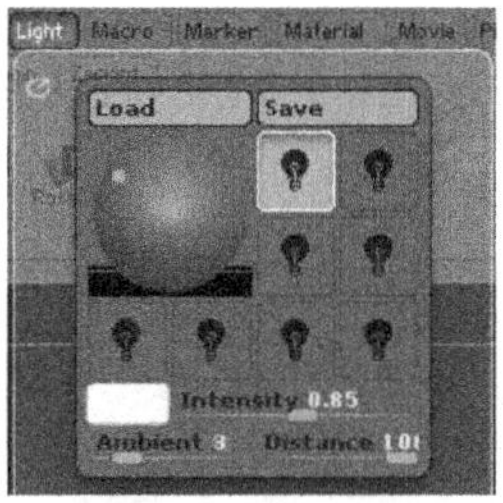

Figure 12-6 *The light positioned in the **Light Placement** window*

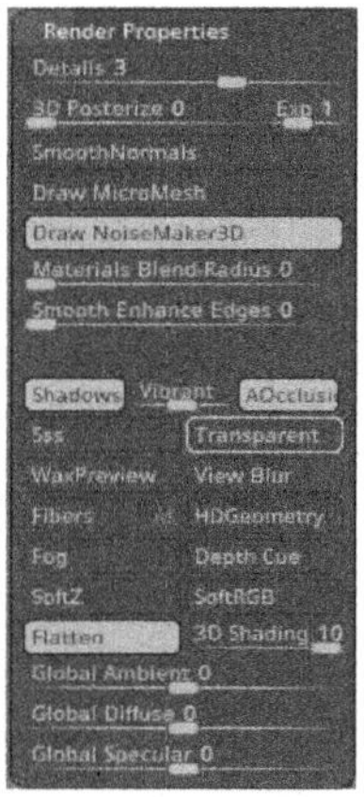

Figure 12-7 *The **AOcclusion** button chosen*

4. Expand the **BPR RenderPass** subpalette in the **Render** palette. In this subpalette, choose the **BPR** button; the scene is rendered and different render passes are generated, refer to Figure 12-8.

Figure 12-8 *The render passes generated*

5. Click on the **Shaded** render pass; the **Export Image** dialog box is displayed, refer to Figure 12-9. In this dialog box, browse to the desired location and create a new folder in that location. Save the **Shaded** render pass in that folder in the .BMP format. Similarly, save the rest of the render passes in the folder.

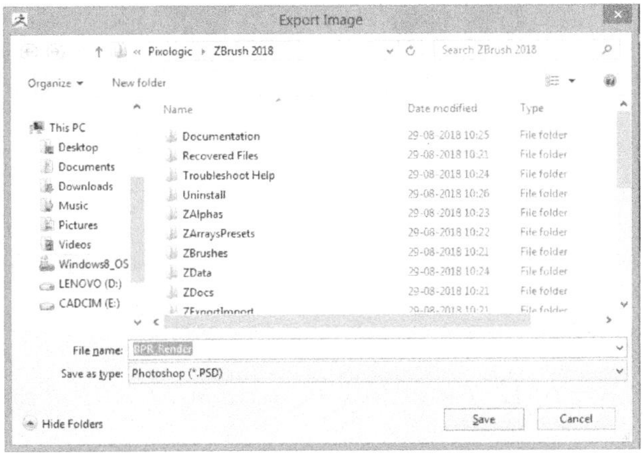

Figure 12-9 *The **Export Image** dialog box*

6. Expand the **Light** palette. In this palette, select the orange dot on the **Light Placement** window and drag it to the position shown in Figure 12-10.

7. Expand the **BPR RenderPass** subpalette in the **Render** palette. In this subpalette, choose the **BPR** button; the scene is rendered and different render passes are generated again.

8. Click on the **Shaded** render pass, refer to Figure 12-11; the **Export Image** dialog box is displayed. In this dialog box, save the **Shaded** render pass in the .BMP format and name it as specular, refer to Figure 12-12.

After generating all the render passes, you can composite these passes in different compositing software applications such as Adobe Photoshop, Nuke, Fusion, and so on.

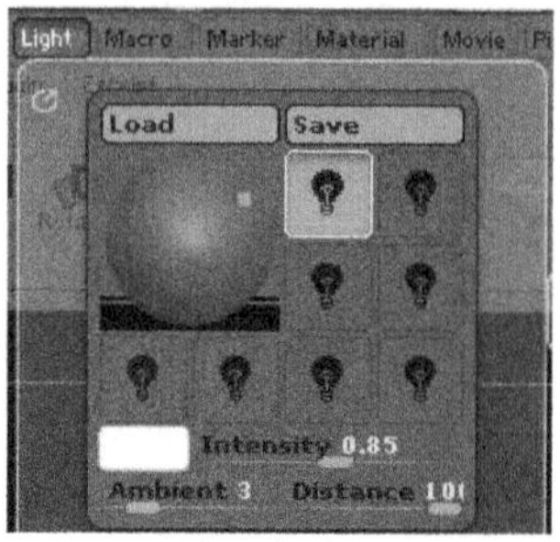

***Figure 12-10** The light positioned in the **Light Placement** window*

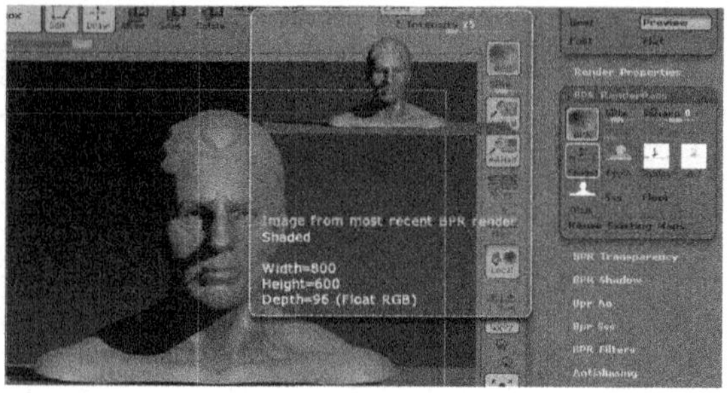

***Figure 12-11** The **Shaded** render pass selected*

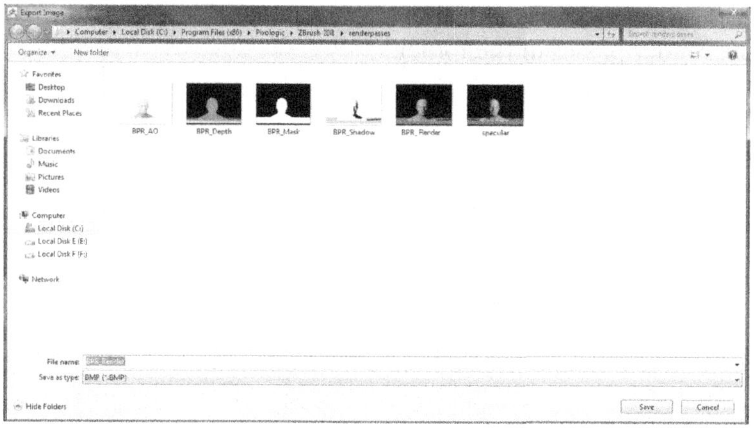

***Figure 12-12** The render passes saved in the .BMP format*

Saving the Scene

In this section, you will save the file using the steps given next.

1. Choose the **Save As** button from the **Tool** palette, the **Save ZTool** dialog box is displayed. In this dialog box, browse to the location *\Documents\ZBrushprojects\c12*.

2. Enter **c12tut1** in the **File name** edit box and then choose the **Save** button.

Tutorial 2

In this tutorial, you will set up a scene and create a KeyShot render of the scene.

(Expected time: 15 min)

The following steps are required to complete this tutorial:

a. Set up the scene.
b. Create the KeyShot render of the scene.
c. Save the scene.

Setting up the Scene

In this section, you will set up a scene.

1. Choose the **Init ZBrush** button from the **Preferences** palette; ZBrush is initialized to its default state.

2. Choose the **Load Tool** button from the **Tool** palette; the **Load ZTool** dialog box is displayed. In this dialog box, browse to*\Documents\ZBrushprojects\c12* and select the **c12tut2start.ZTL** file from it. Next, choose the **Open** button from this dialog box; the model is displayed in the canvas, as shown in Figure 12-13.

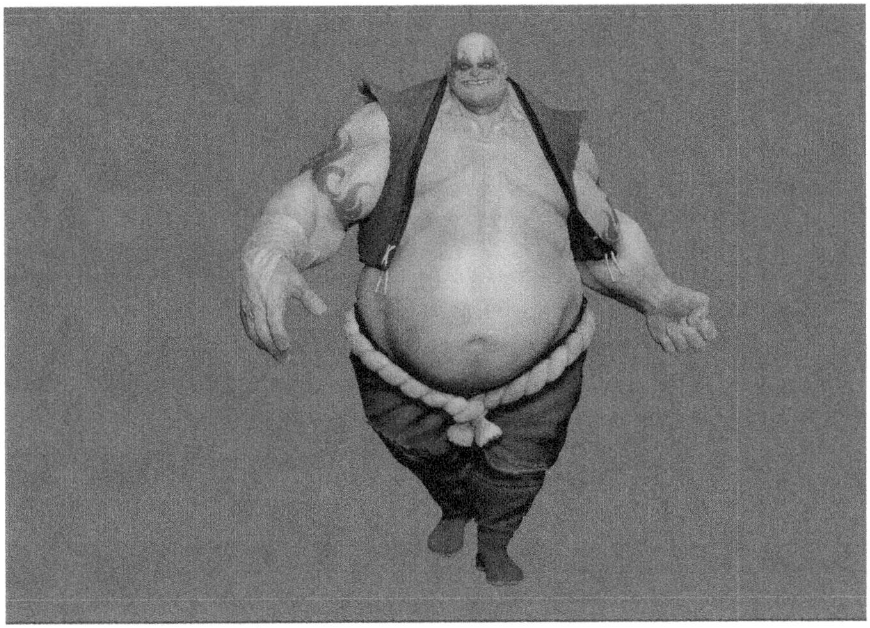

Figure 12-13 *The* ***Kotelnikoff Earthquake*** *model displayed in the canvas*

Create the KeyShot Render of the Scene

In this section, you will create the keyshot render of the scene.

1. Run the KeyShot version available in your system.

2. Expand the **External Renderer** subpalette in the **Render** palette. In this subpalette, choose the **Keyshot** button and press SHIFT+R; the scene is rendered using the KeyShot renderer.

3. Switch to KeyShot renderer and check the render of the model, as shown in Figure 12-14.

Figure 12-14 *The rendered model in the KeyShot renderer*

4. In KeyShot, choose the **Tumble Camera** button from the Ribbon, as shown in Figure 12-15 and rotate the camera to set the model, refer to Figure 12-16.

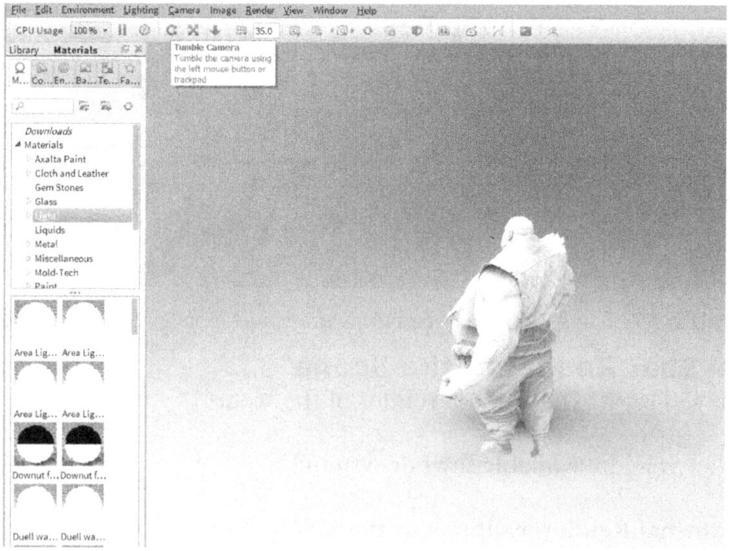

Figure 12-15 *Choose the Tumble Camera from the Ribbon*

Figure 12-16 *Model rotated using the **Tumble Camera** button*

5. Choose the **Material** tab from the **Library** window in the KeyShot renderer, refer to Figure 12-17. Next, expand the **Metal** node and then select and drag the **Anodized Light Blue** material to the dress of the model; this material is applied on the dress of the model, as shown in Figure 12-18.

6. From the **Materials** tab of the **Library** window, drag the **Anodized Rough Black** material to the shoes of the model; the selected material is applied on the shoes of the model.

7. From the **Materials** tab of the **Library** window, drag the **Anodized Brushed Yellow** material to the rope of the model; the selected material is applied on the rope of the model.

8. Choose the **Colors** tab from the **Library** window, then drag the **Sunburst** material to the body of the model; the material is applied to the body of the model. Figure 12-19 is shows the applied materials to the model.

9. Choose the **Backplates** tab from the **Library** window and then drag the **Doch-Design_Dirt-Lot.jpg** material to the canvas; the material is applied to the environment, refer to Figure 12-20.

10. Choose the **Render** menu from the menu bar; a flyout is displayed. Next, choose the **Render** option from the flyout; the **Render** dialog box is displayed. Now, enter the name **scene** in the **Name** text box, as

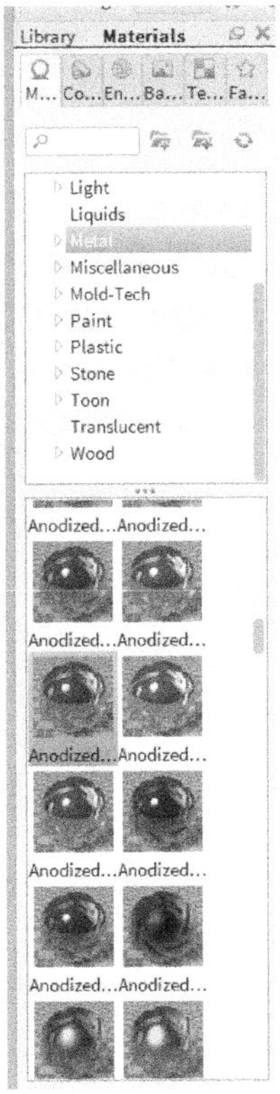

Figure 12-17 Metal *node from Library*

shown in Figure 12-21. Now, choose the **Render** button; the rendering window is displayed, as shown in Figure 12-22.

Figure 12-18 *The **Anodized Light Blue** material applied to the dress*

Figure 12-19 *The **Sunburst** material applied to the model*

*Figure 12-20 The **Doch-Design_Dirt-Lot** material applied*

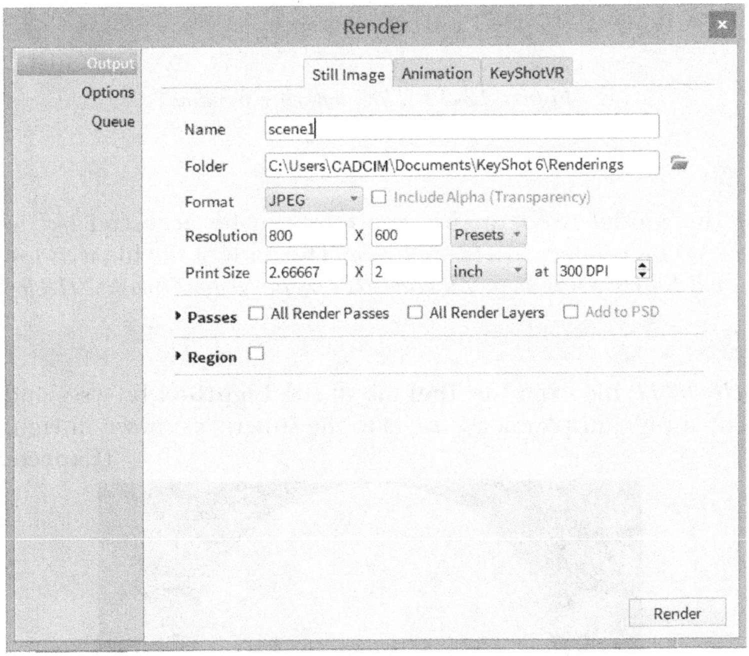

*Figure 12-21 The **Render** dialog box*

Figure 12-22 *The rendering window*

EXERCISE

The output of the model used in this exercise can be accessed by downloading the *c12_ZBrush_2018_exr.zip* from *www.cadsofttech.com*. The path of the file is as follows: *Textbooks > Animation and Visual Effects > Pixologic ZBrush 2018 > Pixologic ZBrush 2018 for Novices.*

Exercise 1

Load the *DemoHead.ZTL* file from the **Tool** tab of the **LightBox** browser and then append a sphere to it. Next, apply transparency settings to the sphere, as shown in Figure 12-23.

(Expected time: 20 min)

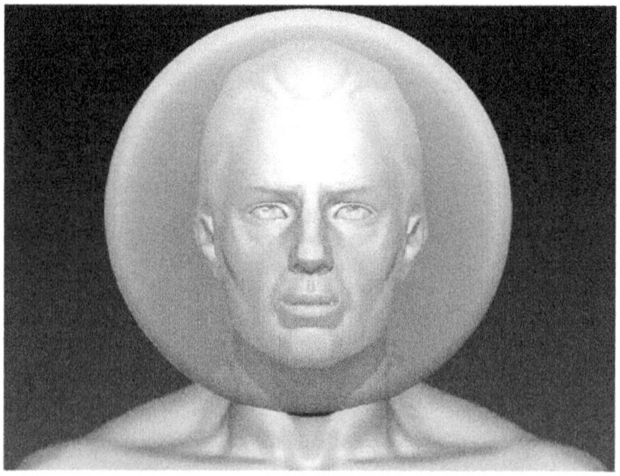

Figure 12-23 *The final output*

Index

Made in the USA
Monee, IL
08 November 2026

67816256R00109